Autorenkollektiv

Physical and Inorganic Chemistry

Resonant Photoacoustic Spectroscopy

P. Hess

Non-Aqueous Electrolyte Solutions in Chemistry and Modern Technology

J. Barthel, H. J. Gores, G. Schmeer, R. Wachter

Subhalides of Tellurium

R. Kniep, A. Rabenau

AKADEMIE-VERLAG · BERLIN

Autorenkollektiv

Physical and Inorganic Chemistry

Autorenkollektiv

PHYSICAL AND INORGANIC CHEMISTRY

With 89 Figures and 26 Tables

AKADEMIE-VERLAG · BERLIN
1983

Die Originalausgabe erscheint im Springer-Verlag Berlin · Heidelberg · New York

Vertrieb ausschließlich für die DDR und die sozialistischen Länder

Erschienen im Akademie-Verlag, DDR-1086 Berlin, Leipziger Straße 3—4
Lizenznummer: 202 · 100/526/83
Gesamtherstellung: VEB Druckerei „Thomas Müntzer“, 5820 Bad Langensalza
Umschlaggestaltung: Karl Salzbrunn
Bestellnummer: 763 251 5 (6773) · LSV 1215, 1255
Printed in GDR
DDR 92,— M

Table of Contents

Resonant Photoacoustic Spectroscopy

Peter Hess

Physikalisch-Chemisches Institut der Universität Heidelberg, Neuenheimer Feld 253,
D-6900 Heidelberg 1, FRG

Table of Contents

1 Introduction

1.1 History

In 1880 Bell discovered the photoacoustic effect in solids [1]. During his experiments with the photophone he noticed that a rapidly interrupted beam of sunlight focused on a solid substance produces an audible sound (see Fig. 1). Within a year three more papers on photoacoustics were published wherein Bell [2], Tyndall [3], and Röntgen [4] described further experiments. The effect was not confined to solids but could also be detected in liquids and gases. In his second paper, Bell described for the first time the resonant photoacoustic effect: "When the beam was thrown into a resonator, the interior of which had been smoked over a lamp, most curious alternations of sound were observed. The interrupting disk was set rotating at a high rate of speed and was then allowed to come gradually to rest. An extremely musical tone was at first heard, which gradually fell in pitch as the rate of interruption grew less. The loudness of the sound produced varied in the most interesting manner. Minor reinforcements were constantly occurring, which became more and more marked as the true pitch of the resonator was neared. When at last the frequency of interruption corresponded to the frequency of the fundamental of the resonator, the sound produced was so loud that it might have been heard by an audience of hundreds of people." (Ref. [2] p. 513–514) Oddly enough, his clear description of the resonance enhancement of the photoacoustic effect did not initiate further studies in this direction. Undoubtedly, progress toward useful applications was inhibited by the limitations set by the light sources and sound detectors used in these early experiments (the human ear was used as detector for rapid periodic pressure changes). Scientists soon lost interest in this effect and no further photoacoustic investigations were performed for half a century.

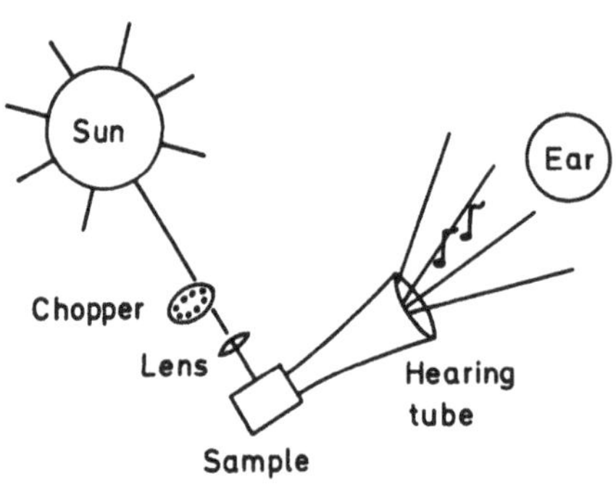

Fig. 1. Historical setup used by Bell. As light sources the sun (or a conventional radiation source) was employed. The acoustic signal was detected with a hearing tube and the ear

A significant step forward was achieved by Viengerov in 1938 [5]. He used blackbody infrared sources, (such as Nernst glowers) for radiation input and a microphone to detect the acoustic signal. His goal was the *analysis of gas mixtures* with a photoacoustic device. The sensitivity of the analysis was limited by the background noise which is produced by absorption of radiation in the cell windows and walls. To minimize these disturbing effects Luft (1943) used two photoacoustic cells, one containing the gas mixture to be analysed and the other the mixture without the substance of interest [6]. With this setup for differential measurements, he could increase the sensitivity of concentration determinations by about three orders of magnitude into the ppm region.

In 1946, Gorelik proposed the investigation of energy transfer processes in gases by phase shift experiments [7]. The idea was that the time delay observable between excitation with the radiation and generation of the pressure signal contains information on energy exchange processes caused in the molecule by collisions. Only two years later the first relaxation measurements were reported by Slobodskaya [8]. During this period the activities were confined to gases and no resonance experiments were performed.

The next important step forward in the development of photoacoustics was the first use of *a laser as radiation source* by Kerr et al. in 1968 [9]. The introduction of laser light sources delivering collimated and monochromatic beams with a high quantum flux opened up completely new areas of research. With this light source an unprecedented sensitivity and selectivity could be achieved from the infrared to the ultraviolet spectral region. Numerous researchers recognized the new experimental possibilities and photoacoustic research began in a variety of areas. An important application was trace gas analysis after Kreuzers pioneering work in 1971 [10]. To improve the detection of atmospheric pollutants Dewey et al. used an acoustic resonance chamber (1973) [11]. With this acoustic amplifier, amplification factors exceeding 100 were attained in these first resonance experiments [11]. Other applications of resonant photoacoustic spectroscopy followed as discussed in detail in this review. Fig. 2 illustrates the change in the setup after about 100 years of photoacoustic research. The progress achieved in the last decade is mainly due to the development of better radiation sources (e.g. lasers) and considerable improvements in the detection system (microphones and electronics).

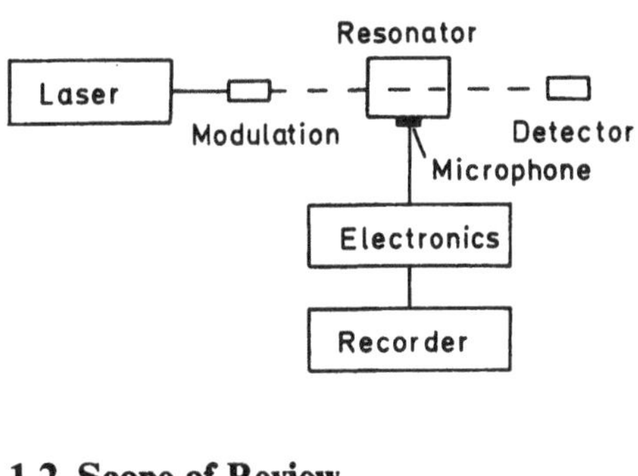

Fig. 2. Modern setup. As light source a cw or pulsed laser is employed. A continuous light beam is chopped by a modulation system of high precision. The acoustic signal is detected by a sensitive microphone and processed by a lock-in analyzer

1.2 Scope of Review

In this article, experiments with gases employing a resonant photoacoustic cell are reviewed. In the next chapter the physical processes occurring in an acoustic resonator are analysed. The third chapter describes experimental setups and components used in resonant photoacoustic spectroscopy. In chapter four the theory of acoustic resonators is outlined and the loss mechanisms which dampen the standing wave in the resonator are discussed. The last chapter gives a detailed discussion of the various applications of resonant photoacoustic spectroscopy known today. These applications are concentration measurements and trace gas analysis, accurate determination of thermophysical properties, and the detection of dynamic processes such as mixing of gases, or chemical reactions. Recent investigations of relaxation processes, spectroscopic experiments, and results obtained for aerosols are also reviewed.

Resonant photoacoustic spectroscopy is only one area of photoacoustics where considerable progress has been made in the last decade. An impression of the large spectrum of recent photoacoustic research can be obtained from the two monographs available on this topic. The first book is entitled: "Optoacoustic spectroscopy and detection" and gives a collection of excellent review articles on fundamental aspects of photoacoustic spectroscopy [12]. In the second monograph, "Photoacoustics and photoacoustic spectroscopy", the entire field of photoacoustics is treated [13]. This book also covers recent advances in the photoacoustic investigation of condensed media such as depth-profiling and photoacoustic microscopy.

2 The Photoacoustic Effect

2.1 Energy Transfer and Radiation Processes

The extremely narrowband emission of a laser allows the specific excitation of molecular states. The non-Boltzmann distribution produced by the excitation process is quickly destroyed by radiation processes and collisional deactivation. The relative contribution of these different deactivation channels depends on the nature of the level excited as shown in Fig. 3. In the microwave region where rotational levels are excited, the radiative life time is very long compared to the very efficient rotational relaxation processes (R—R: rotation—rotation transfer and R—T: rotation—translation transfer). Therefore, the absorbed radiation energy is transformed within a few gas kinetic collisions into translational energy. The situation is similar for

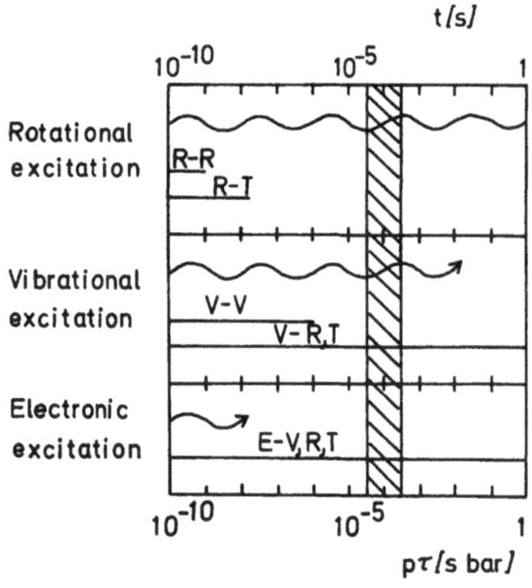

Fig. 3. Deactivation processes and response time of a photoacoustic setup. (wavy arrow) radiative life time in [s], (scale on top). —— range of collisional deactivation by relaxation processes in [s bar], (scale at the bottom); shadowed area indicates the typical response time of a photoacoustic system (sound propagation and microphone response)

excitation of vibrational levels in the infrared region. The radiative life time of vibrational levels still is relatively long (0,01–1 s) and, therefore, deactivation of excited vibrational states by collisional energy exchange is the important channel for restoring equilibrium, at least at higher pressures. Thus, excitation of vibrational levels also leads to an efficient transformation of photon energy into kinetic energy. This can be different for electronic excitation, because electronic levels possess a much shorter radiative life time as can be seen in Fig. 3. The efficiency for energy exchange between electronic states and the other degrees of freedom (E–V,R,T transfer) may vary considerably from system to system. Thus, only part of the photon energy absorbed, is transformed into translational energy in this case. In addition, photons may induce chemical reactions in this wavelength region.

Most resonant photoacoustic experiments performed up to now used an infrared laser to excite a rotational-vibrational state of the absorbing molecule. Energy exchange processes between vibrational levels (V—V: vibration to vibration transfer) and from vibrational states to rotational and translational degrees of freedom (V—R,T transfer) have been extensively studied by several methods. Most data on V—R,T transfer in polyatomic molecules has been obtained by acoustic methods such as ultrasonic dispersion and absorption [14]. The information on V—V exchange comes mainly from laser-induced fluorescence experiments. With this technique many polyatomic molecules have been studied in recent years, yielding a good understanding of energy flow between vibrational states in vibrationally excited molecules. Polyatomic molecules may be divided into four groups according to their relaxation behavior.

1) V—R,T transfer is slow compared to V—V exchange (e.g. CH_4 [16]). Most small polyatomic molecules belong to this group.

2) The rate of V—V exchange between specific modes is comparable to V—R,T transfer (e.g. CO_2 [17]). Only a few molecules show this kind of relaxation behavior.

3) V—R,T transfer is more efficient than V—V exchange between specific modes (e.g. CD_2Cl_2 [18]). Only a small number of molecules with a special level structure belongs to this group.

4) Both, V—V and V—R,T exchange processes are very efficient (e.g. H_2O [19]). For small molecules this behavior is an exception. However, for larger polyatomic molecules with many vibrational degrees of freedom and low lying states, this effective energy transfer seems to be the rule because relaxation pathways with small energy gaps are available. Typical examples for the different kinds of relaxation behavior are given in Table 1 for illustration. This table shows the number of gas kinetic collisions needed to deactivate the lowest vibrational level of the molecule (V—R,T transfer) and the higher level with the longest collisional life time.

It turns out (see Table 1) that in polyatomic molecules the slowest V—V and V—R,T processes are in the region of 10^4–10^5 gas kinetic collision. From kinetic gas theory we can estimate that at a pressure of 1 bar a molecule performs 10^9–10^{10} collisions per second. This means that at one bar the photon energy is transformed into an acoustic signal in about 10^{-5}–10^{-6} sec. For most polyatomic molecules the signal production is even faster. The time needed by the pressure wave to travel from the laser beam area to the microphone in the acoustic cell is therefore in most cases longer than the vibrational relaxation time. For a distance of

Table 1. Number of gas kinetic collisions needed to deactivate the levels with the longest collisional life times

	V–R, T		V–V		Ref.
	mode	coll.number	mode	coll.number	
CH_4	ν_4	20,000	ν_3	50	[16]
CO_2	ν_2	46,000	ν_3	25,000	[17]
CD_2Cl_2	ν_4	10	ν_3	360	[18]
H_2O	ν_2	3	ν_1, ν_3	9	[19]

a few centimeters this transit time is about 10^{-4} sec. The time delay between excitation and detection of the pressure wave, however, is not only influenced by energy transfer processes and the transit time, but also by the response time of the microphone. The two latter factors normally limit the response time of the gas-microphone system to about 10^{-4} sec or longer (see Fig. 3). This discussion clearly shows that only a careful analysis of phase shift experiments yields reliable vibrational relaxation times [15].

2.2 Acoustic Signal in a Resonator

If a single short laser pulse is employed for excitation of a gas, the pressure wave caused by vibrational relaxation is reflected back and forth until it is completely damped by dissipation processes. Figure 4 shows an example for such a photoacoustic signal observed for excitation of H_2Se in a small cylindrical cell with a single CO_2 laser pulse. To achieve a fast repetition of excitation processes, a cw laser is used and the laser light can be modulated, for example, with a mechanical chopper. This procedure allows a modulated energy input into a gas absorbing the corresponding wavelength. If saturation is avoided the acoustic signal produced by collisional deactivation is proportional to the absorption coefficient of the gas and the intensity of the incident laser radiation. The maximum amplitude achieved for the acoustic signal at steady state may be smaller than the maximum amplitude obtainable in a single pulse experiment. The reason for this effect is destructive interference of acoustic waves from the previous cycles with the waves of the following cycle. Thus, the energy still available in the acoustic chamber from previous cycles is not used to amplify successive acoustic signals but produces noise.

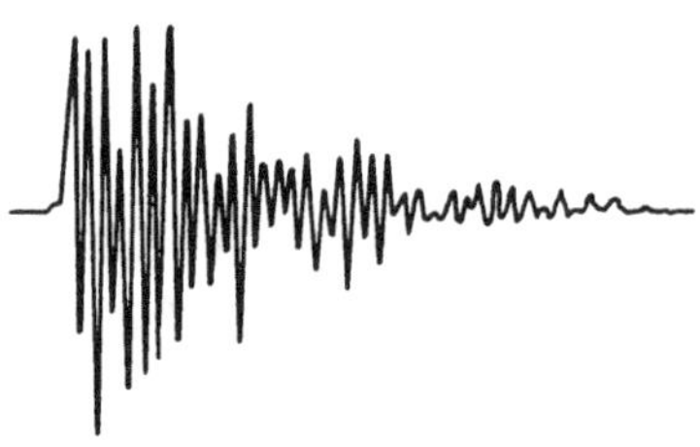

Fig. 4. Typical acoustic signal, observed for vibrational excitation of H_2Se at 60,4 Torr in a small cylindrical cell with a single pulse from a TEA CO_2 laser. The damping of the oscillations is determined by energy dissipation on the walls occurring in the msec range (unpublished results)

In a resonant photoacoustic experiment an acoustic resonator is employed as absorption cell, carefully designed to minimize dissipation of acoustic energy. In addition, the spatial arrangement of the laser beam in the resonator is selected to achieve a minimal disturbance of a standing wave built up in the resonator by successive laser pulses. Finally, the cw laser is chopped with one of the resonance frequencies of the acoustic resonator. This input of photon energy with correct timing leads to the formation of a standing wave in the resonator. Energy from many cycles is accumulated in a standing acoustic wave and the system works as an acoustic amplifier. The final signal amplification obtainable in a resonance system is determined by the resonator losses. After an initial transient state during which energy is accumulated in the standing wave, a steady state is reached in which the energy lost by various dissipation processes is equal to the energy gained by absorption of laser photons.

This discussion shows that special requirements must be met in a resonant photoacoustic experiment. In the next chapter various experimental arrangements are described. In chapter four our present knowledge of dissipation processes is reviewed.

3 Experimental

3.1 Setup for Resonant Photoacoustic Spectroscopy

In Fig. 5 a typical setup for resonant photoacoustic spectroscopy as used in the authors laboratory is shown. The radiation of a cw laser is intensity modulated by a mechanical chopper of high precision. An electro-optic modulation device may also be employed or the laser beam is modulated directly by modulation of its power supply. As already discussed, vibrational excitation with an IR laser, for example, causes a modulated pressure change in the resonator via fast vibrational relaxation. This acoustic signal is detected with a microphone, because these devices provide the highest sensitivity. Detectors employed in calorimetry to measure the heating of a sample such as thermistors or thermophiles are less sensitive and possess a slower rise

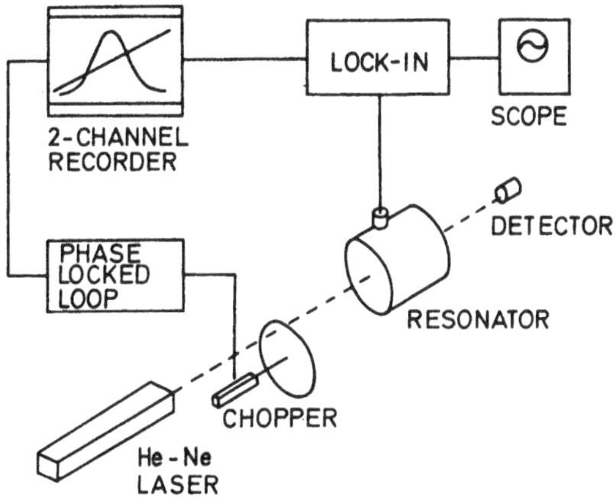

Fig. 5. Setup for resonant photoacoustic spectroscopy with a cylindrical resonator, a He—Ne laser, and simultaneous recording of the chopping frequency variation and the acoustic signal

time. For resonant photoacoustic spectroscopy, electret microphones with a diameter of a few millimeters are utilized by most investigators to minimize the perturbation of the resonator geometry. This will be discussed in more detail in the next section.

The next step is preamplification of the microphone signal. In the case of an electret microphone, this often is performed with a built-in FET preamplifier. For final signal processing a lock-in analyzer is standard today. We wish to point out, however, that resonant photoacoustic experiments can be performed without a lock-in amplifier by applying the simple differential method suggested by Luft [6] with two identical cells, one placed behind the other [20]. The effect of loss of beam power in the second cell can be reduced by sending the laser beam back again through the two cells with a mirror behind the second cell. With this arrangement, signals occurring simultaneously in both cells, such as external noise and window effects, can be suppressed effectively and a simple amplification of the difference signal gives good results.

To record the resonance curve, the chopper frequency is varied around a selected resonance frequency (e.g. the first radial mode of a cylindrical cell). The frequency variation must be linear and slow enough to maintain steady state conditions in the resonator. A satisfactory chopper control can be obtained with a phaselocked loop arrangement [21]. Both the linear chopper frequency change and the corresponding microphone signal are recorded simultaneously with a two channel recorder. From these plots, the resonance frequency, the halfwidth of the resonance curve etc. can be determined. A more accurate but time consuming method is to construct the resonance curve point by point [22]. In this case, the acoustic signal is measured at different fixed frequencies avoiding eventual problems arising from the slow formation of a steady state standing wave in the resonator and the finite time resolution of the lock-in amplifier.

3.2 Excitation Sources, Resonators, Microphones

The experimental possibilities are mainly determined by the radiation source available. Many resonance experiments have been performed with a small 3,39 μm He—Ne laser [11,20,21,23,24]. Unfortunately, only a limited number of molecules can be vibrationally excited with this laser line [25]. The 1,15 μm radiation of a He—Ne laser is absorbed by a large number of molecules which contain C—H bonds. Despite the much smaller absorption coefficients for higher overtones, resonant photoacoustic experiments can be carried out with a more powerful 1,15 μm He—Ne laser [22]. Resonance experiments have also been performed with a cw CO laser [23,26] and a cw CO_2 laser [27,28,48]. The CO_2 laser is an especially versatile radiation source with many lines in the 9 μm–11 μm region and should play an important role in the further development of this field. For a CO_2 laser, intracavity operation of resonant photoacoustic cells has been reported [29,30]. The ideal laser would be a tunable infrared laser to eliminate the problem of an accidental coincidence between laser line and absorbing molecular state of the species of interest. The tunable laser systems available today are expensive and, therefore, only a few experiments have been reported till now. However, even the relatively low output of an optical parametric oscillator system, for example, would be high enough to perform resonance experiments. It is clear that not only IR lasers but also lasers radiating in the VIS or UV spectral range can be used. To detect trace amounts of SO_2 a frequency-doubled dye laser of 1 mW output was employed for electronic excitation of SO_2 around 300 nm [31].

The acoustic resonator must be constructed with great care. The resonant cavity should possess a nearly ideal geometry. Perturbations caused by windows, gas inlet, and microphone should be carefully minimized. The acoustic resonator is characterized by the quality factor Q, which is defined as the ratio of the resonance frequency to the frequency bandwidth between halfpower points. The amplitude of the microphone signal is $1/\sqrt{2}$ of the maximum amplitude at these points, because the energy of the standing wave is proportional to the square of the induced pressure. The acoustic Q should be independent of the level of excitation. In a high quality resonator, Q's of several thousand may be achieved.

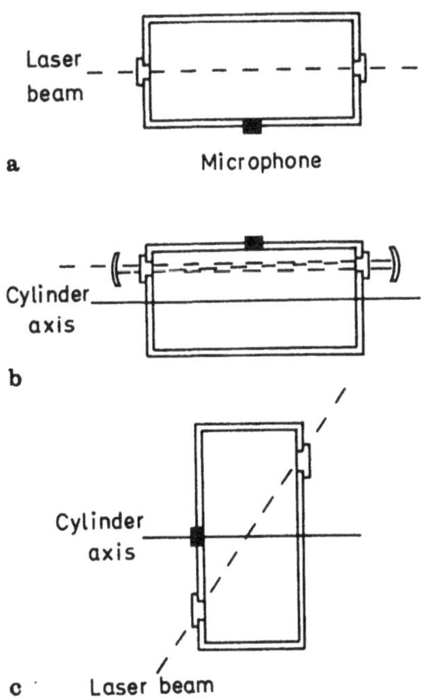

Fig. 6. Different designs used for cylindrical cells. a) Setup for efficient excitation of radial modes in a cylinder. b) Multipass arrangement for excitation of azimuthal modes (see Ref. [31]). c) Cylindrical cell suitable for intracavity operation (see Ref. [30])

Many laser experiments have been performed with cylindrical resonators (see Fig. 6) possessing a size of 10–15 cm in length and diameter [11,20,24]. For these dimensions the fundamental acoustic resonances lie in the kHz region. Increasing the size yields smaller resonance frequencies, but also a decreasing amplitude of the acoustic signal, which is inversely proportional to the cross section of the cell. A multipass arrangement as shown in Fig. 6b with external mirrors to reflect the laser beam back and forth gives considerable improvement for low power excitation or trace gas analysis [31]. Figure 6c shows a Brewster window chamber designed for intracavity operation [30]. This cell is also suited for windowless operation and thus permits, for example, a continuous monitoring of the ambient atmosphere.

For a spherical resonator no resonance experiments with laser excitation have been reported so far. However, acoustic excitation of radial modes with a second microphone was accomplished just recently and, as expected, very high Q factors were obtained in a high quality spherical resonator [32,33]. The goal of these experiments was the measurement of thermophysical properties with high precision. A spherical resonator is also an interesting candidate for photoacoustic investigations.

Resonant photoacoustic studies are normally performed with relatively large acoustic cells possessing resonance frequencies in the 1–10 kHz. The decrease in the acoustic signal caused by the large cell volume and the high frequencies employed is overcome by the high Q factors achieved. In non-resonant photoacoustic experiments much smaller cells are employed and the chopping frequencies applied are usually in the 100 Hz region. For acoustic amplification in this frequency range Helmholtz resonators have been designed. A typical cell is shown in Fig. 7a) where two volumes V_1 and V_2 are connected by a thin cylindrical tube. Helmholtz resonators have been investigated experimentally and theoretically by several groups [34–40]. The resonance frequency can be changed by alteration of the length and the cross-sectional area of the tube, or by changing the volumes V_1 and V_2. These Helmholtz resonators provide a substantial resonance enhancement ($Q \approx 100$) with a relatively small total gas volume of the cell. This type of cell or modifications of this design has been extensively used in photoacoustic spectroscopy of solids [41–43], which is outside the scope of this review. A H-type cylindrical cell with a large volume designed for resonant photoacoustic spectroscopy in gases is shown in Fig. 7b) [48].

As already pointed out, miniature electret microphones are in common use as detectors in resonant photoacoustic spectroscopy [44,45]. Even these small devices may have a pronounced effect on the Q factor of the resonator. For example, the

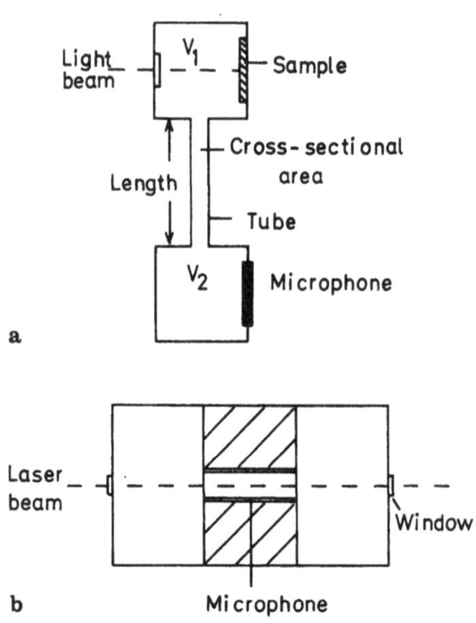

Fig. 7. H-type photoacoustic cells. a) Helmholtz resonator with separated sample chamber and detection chamber for photoacoustic spectroscopy on solids. b) H-type cylindrical cell for resonant photoacoustic measurements in gases. In the tube section a coaxial cylindrical microphone is installed (Ref. [48])

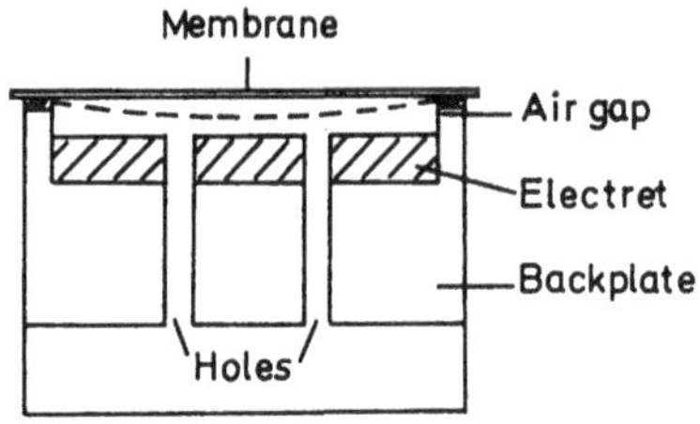

Fig. 8. Cross section of a miniature electret microphone with a separate membrane. This type allows an independent optimization of the membrane and the electret material. Holes to a back chamber decrease the stiffness of the air gap

insertion distance to the axis of a cylindrical cell is critical and must be optimized carefully. Compared with conventional condenser microphones, no external dc polarization voltage is necessary in electret transducers. The electric field is generated by charge trapped permanently in the electret foil which at present consists usually of treated Teflon. In more expensive electret microphones a metallic membrane or a metallized foil membrane is utilized (see Fig. 8). In the most frequently used low cost microphones a thin, permanently charged, movable diaphragm acts as one plate of the capacitor. The frequency response of electret microphones extends beyond 10 kHz and the response to incident pressure waves is linear over many orders of magnitude. Very thin foils can be used as membranes and thus the high sensitivity of about 10 mV/μbar is typical for these devices. A mechanical and an electrical model of the behavior of condenser type microphones is presented by Kreuzer in his contribution to the book edited by Pao [12]. If all sources of noise could be eliminated, the brownian motion of the microphone membrane would determine the ultimate sensitivity of the photoacoustic technique. This noise can be considered as due to pressure fluctuations in the gas as discussed in more detail by Kreuzer [10]. In reality, other background signals such as window absorption limit the ultimate sensitivity.

4 Theory of Acoustic Resonators

4.1 Acoustic Normal Modes in a Spherical and Cylindrical Resonator

The kinetic energy produced by the absorption of a modulated beam of light in a resonator causes the excitation of acoustic normal modes in the cavity. The simple theoretical approach considered in this section is that of a lossless gas in a rigid container with perfectly insulating walls. This simple model yields reasonable values for the eigenfrequencies of the resonator. However, for an accurate description of experimentally determined resonance frequencies and the corresponding Q factors, dissipation effects, deviations of the cavity from the ideal geometry, etc. must be taken into account. The advances made in recent years in an understanding of these loss mechanisms will be discussed in the next section.

The resonator with the highest symmetry is the spherical resonator. A treatment of oscillations in a spherical cavity was already given by Rayleigh in 1894 [46]. A recent discussion may be found in Ref. [33]. In the simple loss-free model, sound is described by the homogeneous wave equation:

$$(\nabla^2 + k^2)\, p(r) = 0 \qquad (1)$$

with $k = \omega/v$, where ω is the angular frequency, v the sound velocity, and p the acoustic pressure. In spherical coordinates, the solutions p(r) have the form of a product of spherical Bessel functions of mth order $J_m(kr)$ and a spherical harmonic. The normal mode solutions of the homogeneous wave equation are determined by the boundary conditions. For rigid resonator walls, the acoustic velocity component normal to the walls must vanish at $r = r_0$, where r_0 is the radius of the sphere. Since the gradient of p(r) is proportional to the acoustic velocity, the gradient of p(r) normal to the boundary must vanish at the boundary:

$$\frac{dJ_m(kr)}{dr} = 0 \quad \text{for} \quad r = r_0 \tag{2}$$

In the following only radial modes of the spherical cavity are considered for which $m = 0$. With the solution

$$J_0(kr) = \frac{\sin(kr)}{kr} \tag{3}$$

we obtain from equation (2):

$$\tan(kr_0) = kr_0 \tag{4}$$

Table 2. Eigenvalues of the radial modes in a spherical resonator

n	a_{on}
1	4,49341
2	7,72525
3	10,90412
4	14,06619
5	17,22076

The index n is used to label the successive roots of this equation. The values of the first roots a_{on} are listed in Table 2. The eigenfrequencies of the radial modes of a spherical resonator can be calculated with these a_{on} values using the equation:

$$\nu_n = \frac{v a_{on}}{2\pi r_0} \tag{5}$$

Now we consider the wave equation for the propagation of sound in a cylinder of radius r_0 and length 1. For cylindrical coordinates (r, φ, z) we obtain the homogeneous wave equation:

$$\frac{1}{r}\frac{\partial}{\partial r}\left(r\frac{\partial p}{\partial r}\right) + \frac{1}{r^2}\frac{\partial^2 p}{\partial \varphi^2} + \frac{\partial^2 p}{\partial z^2} + k^2 p = 0 \tag{6}$$

The solution of this equation in terms of Besselfunctions can be found in Ref. [47] and [48]. The normal mode solutions p_j referring to the resonant frequencies ω_j are determined by the boundary conditions imposed by the rigid cavity. The gradient of the acoustic pressure p normal to the container walls must vanish at the walls:

$$\left(\frac{\partial p_j}{\partial z}\right)_{z=0,l} = 0 \qquad \left(\frac{\partial p_j}{\partial r}\right)_{r=r_0} = 0 \tag{7}$$

These conditions determine the allowed values of $k_j = \omega_j/v$.
If the end walls of the cylinder are at $z = 0$ and $z = l$ we obtain:

$$k_z = \left(\frac{\pi}{l}\right) n_z \qquad n_z = 0, 1, 2 \ldots \tag{8}$$

Applying the boundary condition to the walls at $r = r_0$ leads to the conditions:

$$\left(\frac{dJ_m(k_r r)}{dr}\right)_{r=r_0} = 0 \tag{9}$$

$$k_r = \frac{\pi \alpha_{mn}}{r_0} \qquad \begin{matrix} n = 0, 1, 2 \ldots \\ m = 0, 1, 2 \ldots \end{matrix} \tag{10}$$

where α_{mn} is the nth root of the equation involving the mth order Besselfunction.

Table 3. Values for α_{mn} in a cylindrical cavity

	n = 0	1	2	3
m =				
0	0	1,2197	2,2331	3,2383
1	0,5861	1,6970	2,7140	3,7261
2	0,9722	2,1346	3,1734	4,1923
3	1.3373	2,5513	3,6115	4,6428

The values of α_{mn} for the lowest values of m and n are listed in Table 3. Substitution of the solution p_j into the wave equation (6) gives the resonant frequency ω_j:

$$\omega_j = v(k_z^2 + k_r^2)^{1/2} \tag{11}$$

and this leads to the expression for the frequencies of normal modes in a cylindrical resonator:

$$\nu_{mnn_z} = \frac{v}{2}\left[\left(\frac{\alpha_{mn}}{r_0}\right)^2 + \left(\frac{n_z}{l}\right)^2\right]^{1/2} \tag{12}$$

The pure normal modes are divided into longitudinal, radial and azimuthal modes. Figure 9 gives a graphic representation of these oscillations.

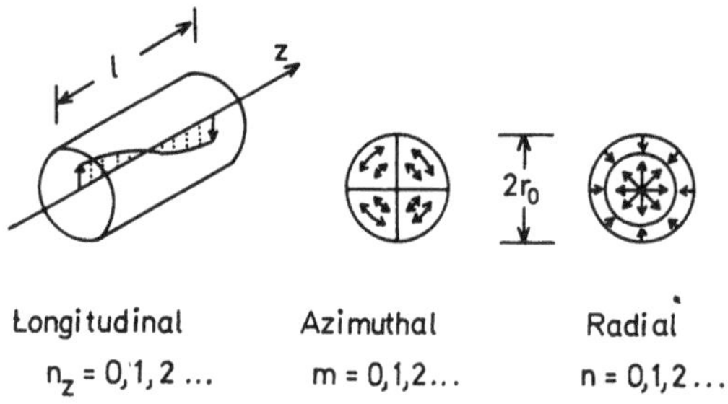

Fig. 9. Acoustic longitudinal, azimuthal, and radial modes in a cylindrical resonator. Only oscillations at the resonant frequencies are amplified and gain significant energy

The integer m determines the azimuthal modes. This number specifies the number of nodal diametral planes in the tube, where the change of the acoustic pressure and the axial velocity are zero. The first azimuthal resonance has been excited with the arrangement shown in Fig. 6b. To gain signal stability, a medium-sized Q factor ($Q \approx 100$) is often favored [31]. The eigenvalue n describes the radial modes and indicates the number of internal nodal cylinders. For $n = 1$, the first radial mode, there is only one nodal cylinder along the tube axis. For this radial oscillation high Q factors ($Q \approx 1000$) have been achieved [20,28]. This is the mode mostly utilized in laser experiments with a cylindrical resonator. Nodal planes and nodal cylinders are absent in the tube for $m = 0$ and $n = 0$. The eigenvalue n_z describes the longitudinal modes as shown in Fig. 9. Such oscillations have been extensively studied employing acoustic excitation with condenser microphones at the end walls of the tube [49,50]. This is one of the important techniques for accurate speed of sound measurements in the gas phase. In such experiments high overtones of longitudinal modes can be excited. Longitudinal modes have also been excited in H-type cylindrical cells [48] and in intracavity operated windowless cells [29].

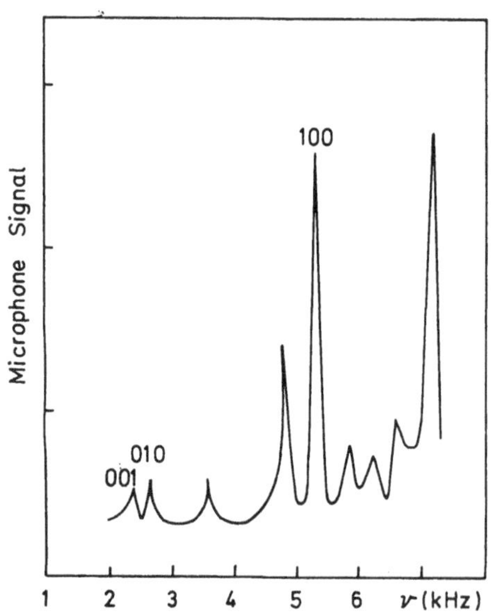

Fig. 10. Photoacoustic resonances measured for a cylindrical cell filled with CH_4. The symbol (nmn_z) is used to label the radial n azimuthal m and longitudinal n_z modes. In addition, several overtones and combinations of these modes can be excited

In Fig. 10 a spectrum of resonances measured for CH_4 in a cylindrical cell with $l = 102$ mm and $r_0 = 51$ mm is shown for illustration. Not only the fundamental modes, but also combinations of these modes are observed. Identification of the resonances can be difficult, if geometric imperfections and dissipation processes lead to larger frequency shifts.

The eigenfrequencies of a spherical or a cylindrical cavity are determined by the size of the resonator and the sound velocity (see eq. (5) and eq. (12)). Therefore, the non-ideal behavior of the gas and relaxation processes causing a dispersion of the sound velocity lead to a pressure dependence of the resonant frequencies. These effects may be a few percent or even more depending on the virial coefficient and relaxing heat capacity. Shifts of the resonant frequencies are also caused by viscous and thermal dissipation. The corrections which must be applied to the ideal model described in this section will be discussed in the following.

4.2 Loss Mechanisms

The energy accumulation attainable in the standing wave of a resonant cavity is many times larger than the radiation energy absorbed per cycle. This acoustic amplification process is limited by various dissipation processes. In order to compare theory and experiment, not only resonant frequencies but also Q factors are measured and compared with model calculations. In recent experiments using high quality spherical and cylindrical resonators, good agreement was obtained with theory.

The first detailed discussion of the various dissipation processes occurring in an acoustic cavity was given by Kamm in 1976 [51]. These losses can be divided into surface effects and volumetric effects. The surface effects are due to the interaction of the standing wave with the internal resonator surface and may be subdivided into the following dissipation processes.

1) Viscous and thermal dissipation inside the boundary layers at the smooth internal surfaces.
2) Losses due to wave scattering at surface obstructions such as gas inlet, microphone, and windows.
3) Compliance of the chamber walls.
4) Dissipation at the microphone diaphragm.

In a carefully designed high quality resonator, the dissipation processes 2), 3), and 4) can be kept negligibly small. It is important to minimize the perturbations caused by these effects, because a theoretical treatment is difficult. It was shown in Ref. [28] and [33] that such an optimization of the cell is in fact possible and then only viscous and thermal boundary layer losses need be taken into account. Throughout the principal portion of the volume of the resonator, the expansion and contraction of the gas occurs adiabatically. Near the walls, however, this process becomes isothermal. This leads to heat conduction, which is responsible for the thermal dissipation process. The viscous dissipation can be explained by the boundary conditions imposed by the walls. At the surface, the tangential component of the acoustic velocity is zero, whereas in the interior of the cavity, it is proportional to the gradient of the acoustic pressure. Thus, viscoelastic dissipation occurs in the transition region.

For purely radial modes in a cylindrical cavity the contribution of these losses to the Q factor is given by [51]:

$$\frac{1}{Q_s} = \frac{1}{l}\left[d_v + (\gamma - 1)\, d_h\left(1 + \frac{1}{r_0}\right)\right] \qquad (13)$$

where $$d_v = \left(\frac{2\eta}{\varrho\omega}\right)^{1/2} \qquad d_h = \left(\frac{2\lambda M}{\varrho\omega C_p}\right)^{1/2} \qquad (14)$$

are the viscous and thermal boundary layer thicknesses, η is the viscosity, λ is the thermal conductivity, M is the mass per mole, C_p is the molar heat capacity, ϱ is the density, and $\gamma = C_p/C_v$.

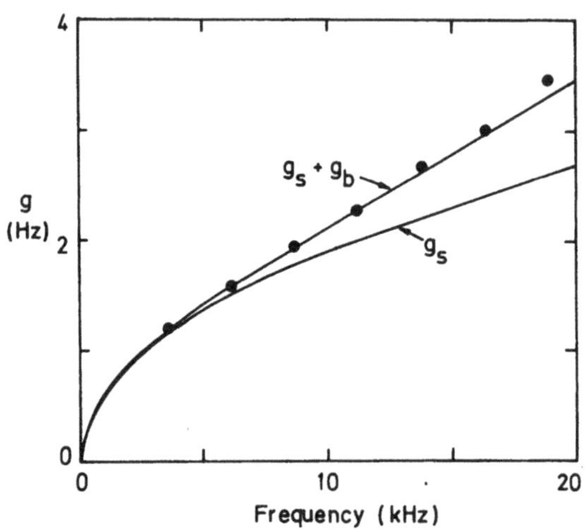

Fig. 11. Comparison of calculated and measured halfwidths of resonances in a sphere filled with Ar at 295 K and 0,05 MPa. The lower curve (g_s) is the calculated contribution of thermal boundary losses. The upper curve ($g_s + g_b$) is in excellent agreement with the experimental points and was obtained by adding the calculated bulk losses to the thermal boundary losses. (Reproduced by permission from Moldover and Mehl [52])

In a spherical resonator no viscous boundary layer losses appear for radial modes. This advantage and the favorable surface-to-volume ratio makes the sphere superior to the cylinder. Very high Q factors (2000–6000) have been observed for carefully constructed devices [32,33]. Figure 11 demonstrates the excellent agreement between experiment and theory obtainable with a spherical resonator. The points show the resonance halfwidths g measured for several overtones of the radial mode in argon at 295 K and 0.05 MPa [52]. The main dissipation comes from the thermal boundary losses g_s. In addition, there is a small loss from free-space viscous and thermal dissipation g_b, which increases with frequency. The frequency shift caused by thermal boundary losses was 0.02% or smaller and the corresponding perturbation due to the bulk losses was negligible [52]. A similar result was achieved with a high quality cylindrical resonator [28]. In the case of the noble gases, the viscous and thermal boundary losses accounted for 90%–100% of the measured losses [28]. In a cylindrical resonator viscous boundary losses make a substantial contribution to the total losses and, therefore, the Q factors are smaller than for a sphere.

The volumetric or bulk losses are caused by the processes which tend to establish equilibrium in the propagating wave. These damping processes are [51]:

1) Free space viscous and thermal losses
2) Relaxation losses
3) Diffusion effects
4) Radiation effects

Friction due to compressional motion and transformation of organized energy into heat due to temperature gradients are responsible for the free space viscous and thermal losses. These two processes are often called Stokes-Kirchhoff losses and play only a minor role as already discussed for monoatomic gases. Diffusion and radiation effects are normally negligible but relaxation effects can make a significant contribution in diatomic and polyatomic molecules. The reason for the relaxational losses is the phase difference between gas pressure and density in the dispersion region, leading to an irreversible conversion of sound energy into thermal energy. This process occurs in a limited pressure region determined by the $p\tau$ values of the gas, where τ is the relaxation time.

The contribution of free space viscous and thermal dissipation and of relaxation damping to the Q factor is given by [53]:

$$\frac{1}{Q_b} = \frac{\omega}{2v^2}\left[\frac{3}{4}\frac{\eta}{\varrho} + (\gamma - 1)\frac{\lambda}{\varrho C_p} + \frac{\eta_{rel}}{\varrho}\right] \tag{15}$$

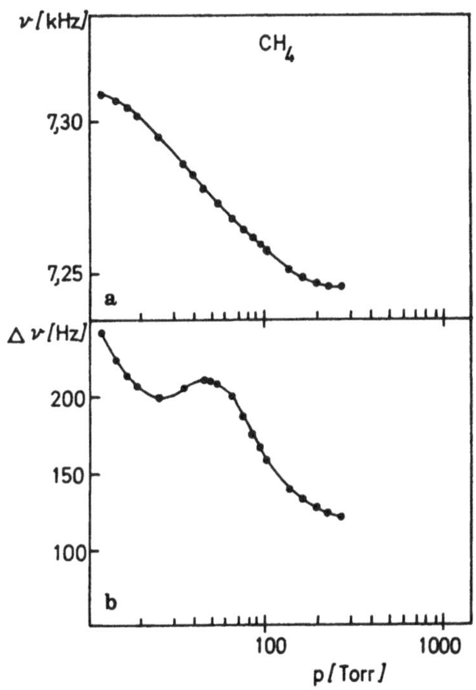

Fig. 12. Effects of relaxation processes. a) Frequency dispersion in CH_4. Due to the small vibrational heat capacity in CH_4, a relatively small frequency shift is obtained. b) Broadening of the resonance curve in the pressure region where relaxation occurs

where η_{rel} is an effective bulk viscosity which takes into account relaxation losses in diatomic and polyatomic molecules. The bulk viscous and thermal losses are small compared to the corresponding boundary layer losses. Only at low pressure can these dissipation processes make a significant contribution. The third term in equation (15), however, is the dominant dissipation process in the pressure region where relaxation processes occur. Figure 12 illustrates the frequency dispersion and the broadening of the resonance halfwidths caused by vibrational relaxation in CH_4 [54]. Accurate measurements for C_2H_4 in a spherical resonator at 300 K and 0,1 MPa show that the relaxation contribution to the resonance width dominates over all other contributions under these conditions [33]. The surface effects are much smaller for C_2H_4 than for Ar, because $\lambda/\varrho C_p$ and $(\gamma - 1)$ are much smaller. It should be pointed out that for excitation of non-radial modes in a sphere or for excitation of resonances in a cylinder, perturbations originating from viscous damping may be more than a factor of 10 higher [55]. Therefore, the excitation of radial modes in a spherical resonator is the best way to study relaxation effects, because the contribution of all other dissipation processes can be kept relatively small. The effect of dispersion of the resonant frequency due to energy transfer processes has been first described for vibrational relaxation in CH_4 employing a cylindrical cell [20]. Such photoacoustic frequency dispersion experiments have also been reported recently by Amer and coworkers for the gases CO_2, N_2O, and SF_6 [28]. In mixtures containing H_2 even rotational relaxation of H_2 could be detected [28].

5 Applications of Resonant Photoacoustic Spectroscopy

5.1 Trace Gas Analysis

As mentioned in the introduction, resonant photoacoustic spectroscopy was introduced by Dewey et al. in 1973, to improve the photoacoustic detection of atmospheric pollutants by acoustic amplification [11]. The characteristics of acoustically resonant and nonresonant cells were compared at different pressure by Gerlach and Amer using a CO laser (≈ 5 mW) to excite CO deluted in N_2 [26]. For the resonant cell the signal fell off linearly with decreasing pressure whereas it remained constant for the nonresonant cell. This indicates that a nonresonant cell is advantageous for low pressure operation. Kamm pointed out that the sensitivity of the photoacoustic analysis can be further increased by a multipass configuration, which increases the photon energy absorbed by the gas considerably in the case of a small absorption coefficient [51]. Such a multipass arrangement (see Fig. 6b) has been used by Koch and Lahmann for the detection of SO_2 concentrations at the 0,1 ppb level [31]. Because background absorption limits the detection of impurities in this case, mirrors outside the resonator and windows with very small losses (antireflection coating on the inner surface) lead to a considerable improvement. The background signal corresponded to an equivalent absorption coefficient of $7{,}5 \cdot 10^{-9}\ \mathrm{cm}^{-1}$. The detection limit for SO_2 ($S/N = 1$) was about 0,1 ppb with a linear increase of the acoustic signal up to 100 ppm. In these experiments SO_2 was excited at ≈ 300 nm with a frequency-doubled dye laser of 1 mW power as mentioned above.

In photoacoustic spectroscopy the signal is proportional to the beam power.

Therefore, the sensitivity can be increased by intracavity operation of the photoacoustic cell. Such a tracegas analyzer with a resonant cell operated in the cavity of a tunable CO_2 laser is described by Shtrikman and Slatkine [29]. The cell could be operated without any windows in the lowest-order longitudinal mode, allowing the determination of the ethylene concentration in the laboratory air as a function of time. Windowless operation leads to additional pickup of ambient acoustic noise. The authors estimate that with a closed cell and a 30 W laser, concentrations lower than 10^{-11} ($3 \cdot 10^{-10}$ cm^{-1}) could be measured [29].

An intracavity photoacoustic detector was also described by Leslie and Trusty [56]. In this work absorption coefficients of CH_4 were measured in the 2500–2800 cm^{-1} region, applying a DF laser for excitation of CH_4. The weakest CH_4 absorption studied was about 10^4 times weaker than the strong CH_4 absorption at 3,39 μm. The intracavity cell was not intended to be resonant in this case, however, by using a resonance found at 310 Hz, a substantially better performance was obtained. The smallest absorption measured with this setup was $7 \cdot 10^{-8}$ cm^{-1} with about 10:1 S/N at a 1 sec time constant.

A photoacoustic cell of novel design was described by Patel and Kerl [57]. This cell consists of a rectangular waveguide with six electret miniature microphones. For excitation of NO a spin-flip Raman laser was used with ≈0,1 W power at 5,3 μm. With this apparatus a NO concentration of $1{,}0 \cdot 10^7$ molecules/cm^3 could be detected. This corresponds to a measurable absorption coefficient of 10^{-10} cm^{-1}. A similar sensitivity may also be obtainable with Nodov's H-type cell (see Fig. 7b) [48].

This discussion shows that a variety of photoacoustic setups have been described in the literature for the determination of small absorption coefficients and the sensitive detection of minute quantities of molecules in gas mixtures. For the reliable analysis of atmospheric or stratospheric constituents, for example, additional requirements must be met. Different species may absorb at the same wavelength, although it is unlikely that several absorption wavelengths coincide. Therefore, a set of reproducible wavelengths should be used for identification and determination of the concentration of species in a multicomponent gas mixture with unknown constituents. Tunable lasers such as diode lasers and spin-flip Raman lasers can be employed to detect interferring species by recording the IR spectrum in a specific spectral region using a suitable photoacoustic cell. In practise, however, most investigations are performed with a CO or CO_2 laser which provide many lines in the 5 μm and 10 μm region respectively.

5.2 Measurement of Thermophysical Properties

The resonant frequencies of an acoustic resonator are determined by the dimension of the cell and by the physical properties of the gas sample determining the sound velocity such as molecular mass, heat capacity, and virial coefficient.

To demonstrate the accuracy of resonant photoacoustic spectroscopy Fig. 13 shows two resonance curves, one obtained for a mixture of 1% CH_4 + 99% N_2 and the other for a mixture of 1% CH_4 + 99% CO. The total pressure was 700 Torr. The molecular mass and the heat capacities of N_2 (M = 28,01 g/Mol, C_p^0 = 29,12 J/K, Mol) and of CO (M = 28,01 g/Mol, C_p^0 = 29,15 J/K, Mol) are nearly identical and the difference in the virial coefficients is small. Therefore, a similar resonant

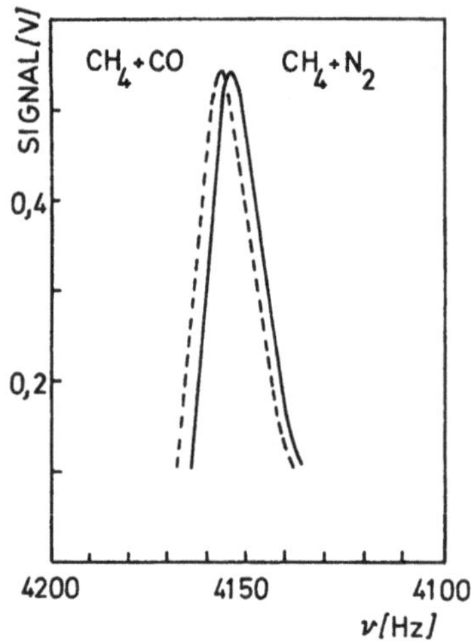

Fig. 13. Independent measurement of a resonance frequency in two mixtures with nearly identical mean masses and heat capacities. The deviation of 3 Hz may be due to inaccurate preparation of the mixtures

frequency is expected for these two mixtures. In fact, the measured frequencies for the first order radial mode in a cylindrical cell are 4155 Hz for CH_4—N_2 and 4158 Hz for CH_4—CO [58]. This indicates an accuracy better than one part per thousand. The main error in this experiment is probably due to deviations from the intended composition of the mixtures.

An example for the accurate determination of sound velocities and heat capacities is given by Thomas III et al. [24]. Using a cylindrical resonator, they investigated mixtures of 99,1% $^{12}CO_2$ + 0,9% CH_4 and of 99,1% $^{13}CO_2$ + 0,9% CH_4 at 760 Torr. From the resonance curves shown in Fig. 14 the authors deduced the first experimentally determined values of $\gamma = 1{,}279$ and $v = 264{,}4$ m/sec for $^{13}CO_2$ at 295.6 K [24]. The shift of the resonant frequency from 3032 Hz to 2994 Hz due to the mass difference of the two isotopes is in agreement with the prediction of eq. (12). Doping

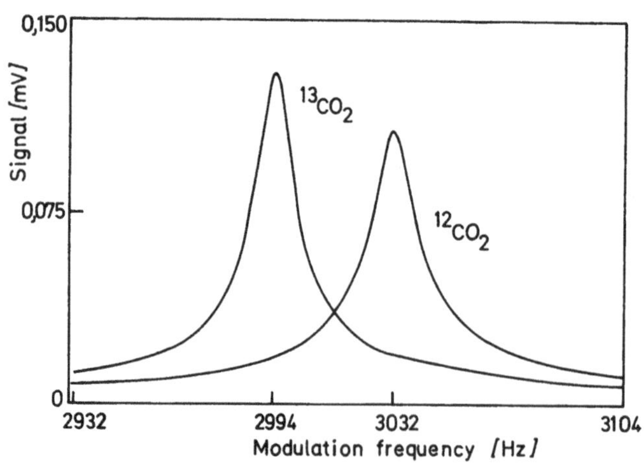

Fig. 14. Frequency of the first radial mode in a mixture of 99,1% $^{12}CO_2$ + 0,9% CH_4 at 760 Torr and a mixture of 99,1% $^{12}CO_2$ + 0,9% CH_4 at 760 Torr and 295,6 K. (Reproduced by permission from Thomas III, Kelly, and Amer [24])

with small amounts of the optically absorbing gas CH_4 makes it possible to investigate gases which do not absorb the 3,39 μm radiation of a He—Ne laser. This doping technique allows the extension of a photoacoustic analysis to any system using a fixed frequency laser source such as an inexpensive He—Ne laser.

Very precise speed of sound measurements were performed by Mehl and Moldover utilizing a spherical resonator and acoustic excitation of radial modes [33,52,55]. As already discussed, the frequencies of radially symmetric resonances in a sphere seem to be quite insensitive to details in the resonator construction. We also assume that the results do not depend on the methods used for excitation and detection of the resonances. The ratio of the speed of sound of a test gas and the speed of sound of a reference gas can be determined with a precision of a few parts in 10^6 by the measurement of the frequencies and widths of radial resonances in a sphere [55]. Monoatomic gases such as Ar are preferable as reference gas. In the limit of zero frequency and zero pressure the sound velocity determined for Ar agrees with the value predicted from kinetic gas theory:

$$v = \left(\frac{5RT}{3M}\right)^{1/2}$$

within the accuracy with which the diameter of the spherical resonator is known, namely $\pm 0{,}01\,\%$ [33].

As test gas C_2H_4 was studied by Mehl and Moldover extensively [33,55]. The speed of sound data obtained, agree with acoustic data determined by Gammon within the scatter of his data ($\pm 0{,}02\%$ in v^2) [59]. Gammon used for his experiments a conventional variable path acoustic interferometer, where the wavelength of sound is measured by moving the transducer from resonance to resonance. The ideal heat capacity at constant pressure C_p^0 derived by Mehl and Moldover from their and Gammon's data are compared in Fig. 15. In addition, they compare the experimental values with theoretical values determined from spectroscopic data. A set of frequencies of the normal modes reported by Herzberg [60] and a somewhat different set given by Shimanouchi [61] were used to calculate the contribution of the vibrational degrees of freedom to the heat capacity. There was a large discrepancy only for the frequencies of the lowest bending mode of C_2H_4 with $\nu_{10} = 810{,}3\ cm^{-1}$ [60] and $\nu_{10} = 826\ cm^{-1}$ [61]. As can be seen in Fig. 15, Herzbergs value should be

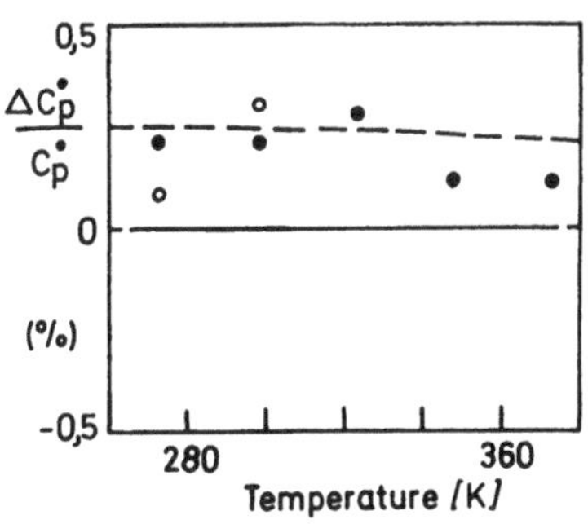

Fig. 15. Comparison of calculated and measured ideal heat capacity of ethylene. Dashed curve: calculated with vibrational frequencies taken from Ref. [60]); Solid curve: calculated with frequencies taken from Ref. [61]); ● experimental results Ref. [55]; ○ experimental results Ref. [59]. (Reproduced by permission from Mehl and Moldover [55])

preferred if the simple model of noninteracting harmonic oscillators is valid. This comparison demonstrates the high accuracy of the acoustic data.

The acoustic virial coefficient of C_2H_4 was determined from the pressure dependence of the sound velocity in the temperature range of 270 K–380 K [55]. The second virial coefficient B(T) was calculated from these acoustic virial coefficients and literature values available for B(T) near 270 K. The resulting temperature dependence and a comparison with data reported by Douslin and Harrison [62] and by Waxman and Davis [63] is shown in Fig. 16. The maximum deviation is 0,8 cm^3/mole at 373 K.

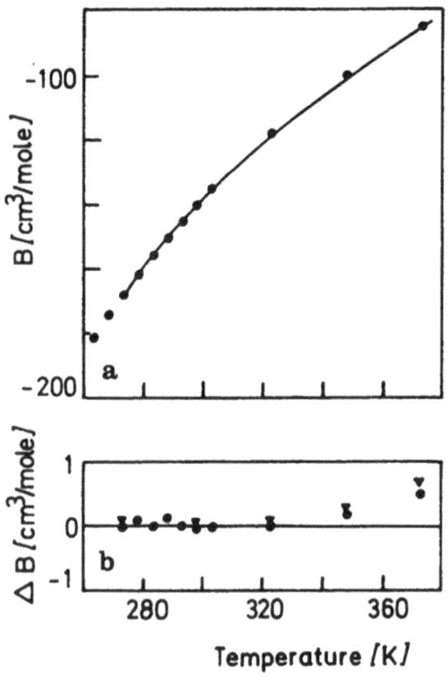

Fig. 16. Virial coefficient of ethylene. a) Solid curve: calculated from experimental results of Ref. [55]; ● experimental data Ref. [62]). b) Comparison of data from Ref. [62]) ● and Ref. [63]) ▼ with the smooth curve of 16a). (Reproduced by permission from Mehl and Moldover [55])

5.3 Detection of Kinetic Processes

In a mixture of ideal gases the sound velocity and consequently the resonant frequencies of a resonator depend on the effective specific heat ratio and the average molecular mass of the mixture. Chemical reaction and mixing processes, for example, are normally accompanied with a change of these properties. Therefore, such dynamic processes can be monitored by a repeated measurement of the shift of one of the eigenfrequencies of the resonator as a function of time.

The time dependence of mixing processes between CH_4 and N_2 has been studied under various conditions by repeated recording of the acoustic resonance curves of the first radial mode in a cylindrical cavity [21]. Figure 17 shows the calculated resonance frequency of this mode as a function of composition. The difference in the resonance frequencies of the pure species is about 1140 Hz and thus, a concentration change of less than 1% can be detected. A scheme of the experimental arrangement is presented in Fig. 18. For fast addition of the second component from a high pressure bottle the measured mixing time varies between several minutes to several hours depending on composition [21]. If any pressure difference is avoided by mixing

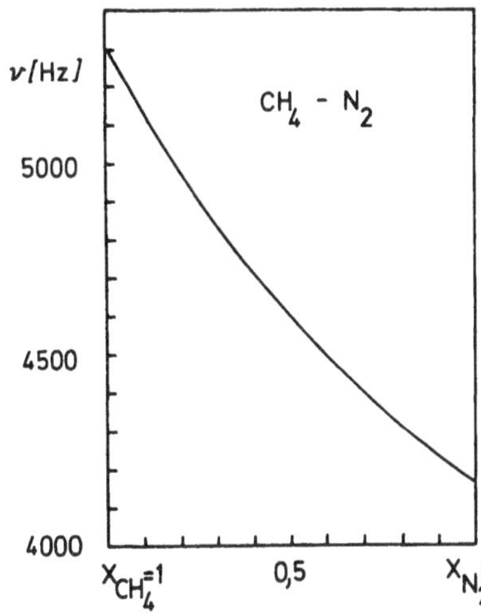

Fig. 17. Calculated resonance frequency of the first radial mode as a function of composition for the system CH_4—N_2 at 294 K

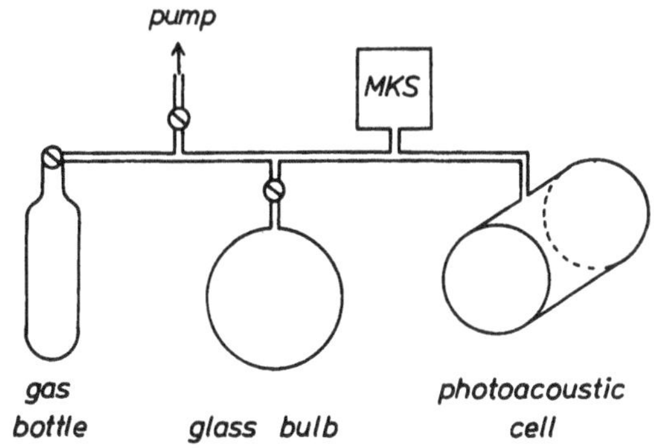

Fig. 18. Scheme of the setup employed to study the kinetics of mixing processes for different mixing procedures

the gases from two connected bulbs with equal pressure, several days may be needed to reach the equilibrium concentration in the resonator. In Fig. 19 the results of such a diffusion controlled mixing experiment are presented. A comparison with Fig. 17 indicates that the resonator was filled with CH_4 at the beginning and the final composition corresponded to 50% CH_4 and 50% N_2. If the two components are condensed together for a short time by cooling with liquid helium, the mixing process

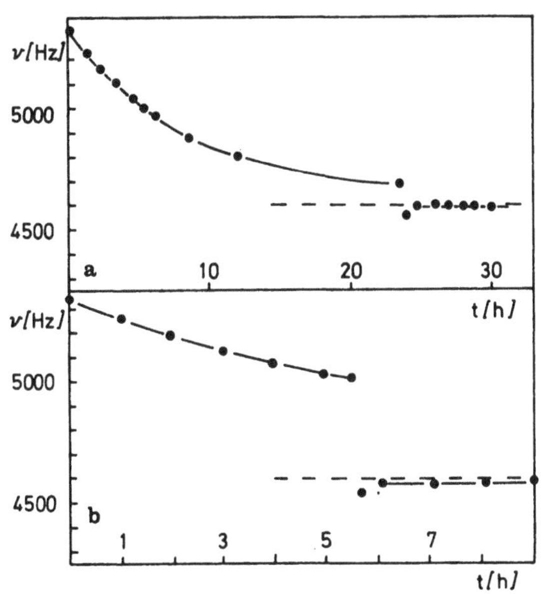

Fig. 19. Diffusion controlled mixing between CH_4 and N_2. The equilibrium concentration (50% CH_4 + 50% N_2) is indicated by the dashed line. a) Freezing after 24 hours. b) Freezing after $5^1/_2$ hours

is accelerated considerably. From Fig. 19 details of processes occurring can be recognized, for example, that after condensation, equilibrium is approached from the N_2 side [21].

This technique for the analysis of dynamic processes can be very accurate if the two components possess distinct masses and heat capacities. The time resolution is limited by the properties of the acoustic resonator and lies in the range of several seconds. To observe faster kinetic processes, a pulse to pulse analysis of the photoacoustic signal in a nonresonant cell can be employed. With such a technique the kinetics of H—D exchange in the system H_2S—D_2S has been investigated by isotope specific vibrational excitation of D_2S and HDS with a TEA CO_2 laser [64]. No influence of the laser radiation on the progress of the exchange reaction was detected. Thus, this method allows the analysis of a chemical reaction by a concentration measurement without perturbation of the reaction process. The ultimate time resolution is determined by the time needed for the generation and detection of the acoustic signal which may be in the msec range in this case.

The excitation of radial and longitudinal modes in a resonant chamber employing a pulsed laser was investigated by Bruce et al. [65]. Single pulse operation leads to a loss of phase information and a reduced sensitivity. To maximize the sensitivity, a unique signal processing scheme was developed by the authors. This sophisticated processing of the acoustic signal allows the effective separation of the signal from broadband noise components. This new technique combining single pulse laser excitation and a resonant cavity has so far only been used to measure the absorption of CH_4 at DF laser wavelengths [65].

5.4 Relaxation Processes

As already indicated, the photoacoustic method can be applied to determine collisional lifetimes of specified quantum states and to study the routes of energy exchange in polyatomic molecules. Mainly two methods have been employed in the last 35 years to monitor the transformation of photon energy absorbed in a defined molecular state to thermal energy by non-radiative processes. These methods measure either the phase lag or the amplitude of the acoustic signal over a suitable frequency region. An information which can only be obtained by these methods is the ratio of energy released fast (e.g. by V—V processes) to the energy released much slower, for example, by V—R,T processes [66]. This piece of information can give additional insight into the complicated energy distribution processes occurring in a polyatomic molecule as shown in Ref. [66] for CH_4 and in Ref. [67] for CH_4 mixtures with inert gases. Because a theoretical modelling of the complicated processes going on in the acoustic cavity is not possible, a calibration of the system is needed. This ideally means a comparable measurement with the same gas under identical conditions but without relaxation effects, or in other words, unmeasurably fast energy exchange. In practise this cannot be realised and, therefore, the calibration error limits the accuracy achievable with these techniques. Often it would be advantageous to work at high chopping frequencies to exclude disturbing effects, but in the high frequency region the sensitivity decreases significantly in a non-resonant cell.

With acoustic resonators, however, experiments are possible in the high frequency region. The high Q-factors attainable when the modulation frequency coincides with one of the acoustic resonance frequencies of the cavity, permit sensitive measurements in the kHz range.

Another advantage of the resonant spectrophone is a uniform heating of the whole volume by the standing wave and not only heating of the beam volume and the surroundings. As pointed out by Frank and Hess the dispersion of the resonance frequency occurring in the relaxation region permits an accurate determination of relaxation times [20]. The highest accuracy is achieved if not only the resonance frequency, but also the Q-factor is included into the analysis. The use of an optimized resonator, where the cavity losses can be calculated using the formulas given above, allows the determination of relaxation data with high precision. Such a procedure is laborious, but no calibration is needed and additional information on transport properties can be obtained. In a recent paper, Zharov and Montanari describe simpler methods for the determination of vibrational relaxation times from resonant photoacoustic experiments [68]. In these simplified methods the acoustic signal is measured only at two resonance frequencies or at one frequency and several pressures. However, a calibration is necessary and the experimental conditions must be selected carefully to minimize the error [68].

5.5 Spectroscopy

As already mentioned, the detection of ultra-low concentrations of pollutants with resonant photoacoustic spectroscopy has been the subject of several theoretical and experimental studies. In these trace gas experiments a high total gas pressure of about 1 bar is used in most cases. For spectroscopic high resolution studies in the gas phase, low pressures must be employed in the acoustic cell. Therefore, the pressure behavior of the photoacoustic detector is of great interest for spectroscopic applications. Kraft and Bevan have compared the performance of a typical small volume non-resonant cell with a typical resonant acoustic cell in the large pressure range of 1–760 Torr [69]. The non-resonant cell had a gas volume of 0,4 cm^3 and the color center laser employed to excite combination bands of CO_2 in the 2,77–2,78 μm range was modulated with chopper frequencies of 50 and 500 Hz. As resonant cell a cylinder with a diameter and length of about 10 cm was used. The strongest resonance observed for CO_2 was the first radial mode at 3301 Hz with a Q-factor of 470 at 760 Torr. A comparison of the acoustic signal and of the signal/noise ratio of these two cells shows that the non-resonant cell is significantly more sensitive than the resonant cell as shown in Fig. 20. At 1 Torr a spectrum could only be observed with the non-resonant cell and above 5 Torr the signal of this cell was about 100 times greater than the signal of the resonant cell. As already mentioned these results were obtained for two typical, not fully optimized acoustic cells. The experiments of Kraft and Bevan clearly show that a standard small volume non-resonant cell poses fewer problems and, therefore, might be preferred in practical spectroscopic applications [69]. The small volume cell is easier to handle, less sensitive to extraneous acoustical noise, and only a small amount of sample is needed. The main advantages of the photoacoustic detection technique over low pressure high resolution absorption

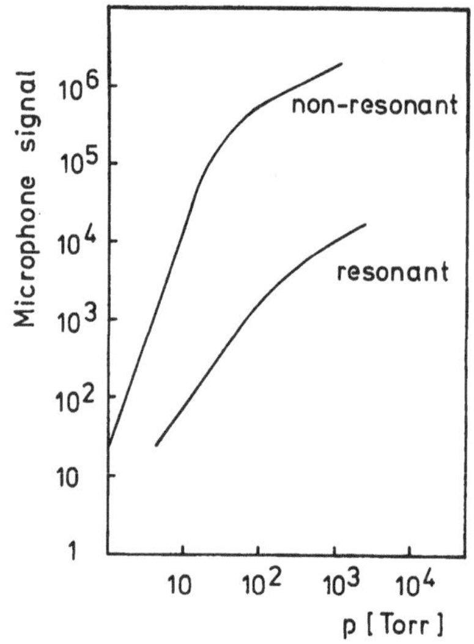

Fig. 20. Comparison of the signal of a resonant and non-resonant cell produced by excitation of CO_2 combination bands with a color center laser [69]; upper curve: signal of the non-resonant cell versus pressure for a modulation frequency of 50 Hz; lower curve: signal of the resonant cell versus pressure for excitation of the first radial mode at 3301 Hz.

studies with lasers are insensitivity to source noise and the absence of background signal.

A high resolution study of the ν_3 band of HCN at 3 μm has been recently reported using a standard small volume cell of non-resonant design [70]. The 25 lowest rotational levels of the ν_3 band were excited with a tunable color center laser at an HCN pressure of 20 m Torr. The pressure of the N_2 buffer gas was varied between 20 and 400 Torr to investigate pressure broadening and to detect possible frequency shifts induced by the buffer gas. In intracavity experiments performed at an HCN pressure of only 0,02 mTorr no shift of the transition frequency could be detected, to the accuracy of the wavemeter readout. The results of this photoacoustic analysis of the ν_3 band of HCN are:

$$\nu_0 = 3311.472 \pm 0.001 \text{ cm}^{-1}$$

$$B' = 1.467797 \pm 0.000001 \text{ cm}^{-1}$$

These values are in good agreement with the accurate spectrographic experiments performed at much higher HCN pressures and calculated values of Rank et al. [71].

The high sensitivity of photoacoustic spectroscopy also allows the investigation of overtones, combination bands, and hot bands possessing a very small absorption cross-section. Such weak transitions cannot be saturated with present cw lasers even by intracavity operation of the photoacoustic cell. The higher laser intensity in the cavity then leads to an enhancement of the acoustic signal, which is proportional to the laser intensity below saturation. For efficient intracavity operation the photoacoustic cells are equipped with wedged Brewster angle windows. Such an arrangement disturbs the geometry of the resonator, unless a special design is used as, for example, the cell shown in Fig. 6c). Nevertheless, Q-factors above 500 have been achieved with the less efficient design using a cylindrical resonator and conventional excitation along the cylinder axis by Smith and Gelfand [72]. In these intracavity experiments a dye laser was used to excite the R(0), R(1), and R(2) transitions of the 5–0 overtone band of HD. The dye laser output was narrowed and scanned across

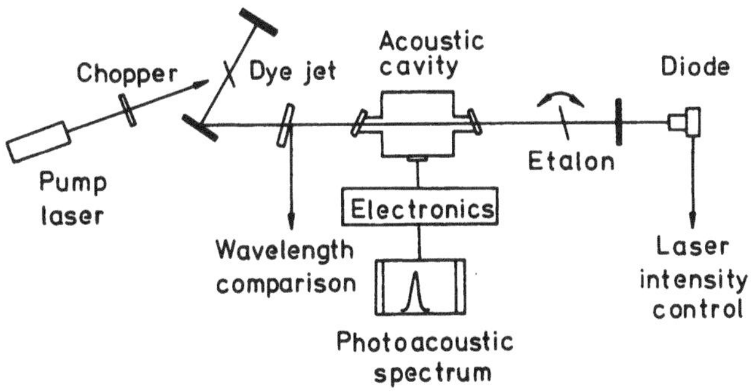

Fig. 21. Scheme of an intracavity photoacoustic spectrometer as described in Ref. [72], the dye laser is pumped with an argon ion laser; the dye laser output is narrowed with an birefringent filter and an etalon; the wavelength is varied by tilting the etalon

the HD absorption profiles. Details of the setup are shown in Fig. 21. The wavelength accuracy achieved was better than 0,02 Å. Within experimental error, the frequencies of the vibration-rotation transitions investigated were in agreement with data given by McKellar, Goetz, and Ramsay [73]. Using the line strength determined by these authors for the 5–0 R(1) transition as a standard, Smith and Gelfand obtained line strengths for the other two transitions 5–0 R(0) and R(2), which were in excellent agreement with the former results. This work demonstrates that the response of a photoacoustic detector can be calibrated by normalization. The remarkable sensitivity of intracavity photoacoustic spectroscopy can be seen from the fact that these experiments were conducted with a pressure-pathlength product of only 10^{-6} of that used by McKellar, Goetz, and Ramsay in the former experiments.

Intracavity dye laser photoacoustic spectroscopy today is a very fruitful area of research. With this ultrasensitive technique, highly forbidden overtone transitions have been studied in more than 50 molecules in the spectral region 1000–4000 nm. The method allows the detection of 9–0 transitions with absorption coefficients as low as 10^{-9} cm^{-1} [74]. Window noise due to scattering and absorption of radiation at the Brewster windows of the sample cell is one of the limiting factors in these intracavity experiments. In molecules containing CH bonds, a highly localized and bond-selective excitation of local modes is possible. The broad overtone bandwidths observed in this case give new insight into the intramolecular dynamics of highly vibrationally excited states in polyatomic molecules [74]. As mentioned before, an enhancement of the acoustic signal can be achieved in the frequency region of acoustic cell resonances. Glass tubes with a small diameter (1–2 cm) possess high

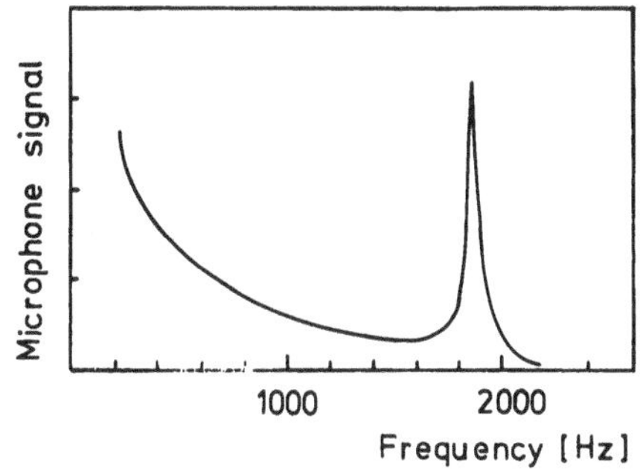

Fig. 22. Dependence of the acoustic signal of an intracavity cell with Brewster windows on the modulation frequency. In a long tube the lowest resonance is the lowest order longitudinal mode, which is excited to enhance the signal [74, 75]

frequency radial and azimuthal modes. The low lying longitudinal modes are distorted by the Brewster windows and, therefore, the Q-factor is normally below 100. Figure 22 gives an example for the enhancement attainable by excitation of those longitudinal modes in comparison to the signal level at low modulation frequencies [75]. The resonance frequency depends on the sample gas. Therefore, the acoustic signal is maximized by tuning the modulation frequency to an optimal value in each system [74].

5.6 Investigation of Atmospheric Aerosols and of Particulate Emission in Combustion Processes

Aerosols in the atmosphere may influence the climate and the temperature of the earth and, therefore, the investigation of light scattering and absorption by aerosols is of considerable importance. The dependence on particle size distributions and optical properties is different for the two processes and this means that different information can be obtained by scattering and absorption measurements. The scattering process has been studied in many systems and is often the main contribution to the extinction of light. On the other hand, the spectrophone seems to be the ideal instrument to study absorption of light by aerosols, because of the insensitivity of the photoacoustic effect to scattered light. The generation of sound waves in the acoustic cell is determined in such an experiment by the rate at which the heat is transferred from the light absorbing particles to the gas. The time constant for this heat transfer is proportional to the square of the particle radius, proportional to the heat capacity per unit volume of the particle, and possesses a reciprocal dependence on the thermal conductivity of the gas [76]. Thus, at high modulation frequencies, no sound is generated by large particles. At 4 kHz for example, the absorption of particles only up to about 3 μm in diameter can be detected.

A resonant spectrophone with holes for *in situ* measurements in the ambient air is described in Ref. [77]. With this cylindrical cell, where the laser beam passed through the cavity near the wall through holes, a Q of about 200 was achieved (see Fig. 23). The acoustic chamber was isolated by several sound barriers; gas could be introduced by a separate gas line if desired through a filter or by pumping on the gas line through the laser ports. The sensitivity of the arrangement was demonstrated by monitoring NO_2 in air. Concentrations less than 0,1 parts in 10^9 could be detected by passing the laser beam four times through the cell using mirrors. These measurements were made with a 0,5 W argon laser at 514,5 nm chopped at 4 kHz.

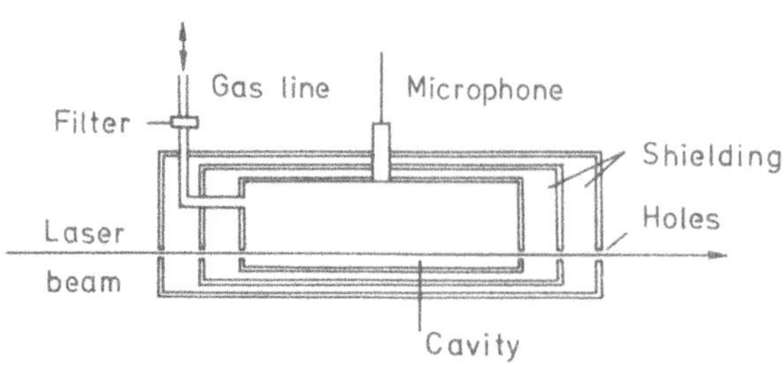

Fig. 23. Scheme of a resonant photoacoustic cell for in situ measurements in the ambient atmosphre [77]. The laser beam enters and leaves the cell through holes: the sound barriers provide an efficient shielding against room noise

In other experiments cigarette smoke was investigated. In aerosols produced by smoke off the tip of the cigarette, absorption is much larger than scattering and the ratio of the contribution from absorption to scattering is about 7. In this case carbonaceous particles are formed with a diameter of ≈0,01 μm as known from pyrolysis experiments. On the way through the cigarette, the particles coagulate into aerosols of greater diameter and, as expected, scattering then increases drastically. In fact the ratio measured for absorption to scattering for such aerosols is only about 0,8 [77]. More experiments on the detection of carbonaceous aerosols with a spectrophone are described in Ref. [78].

The study of carbonaceous particulate originating from flames also finds increasing interest [79,80]. The conventional methods such as filter collection of soot from combustion processes is a long, tedious process, giving no information on fluctuations during the sampling period. The photoacoustic technique, however, provides a "real time" monitor of the particulate. In Ref. [79] simultaneous measurements of the absorption spectrum, the photoacoustic spectrum, and light scatter of soot from a propane-oxygen flame are reported. In these experiments the radiation of a dye laser, which was tunable in the spectral range of 590–625 nm was square-wave modulated at 4 kHz. The resonant acoustic cavity had a Q of about 200. A schematic diagram of the experimental setup is shown in Fig. 24. The soot was produced using a premixed propane-oxygen flame at one atmosphere. As carrier gas N_2 was added, 30 cm downstream from the flame in ten-fold excess. This dilution procedure also minimized water condensation in the acoustic cavity. The absorption and photoacoustic spectra were identical in the spectral range investigated. An absolute calibration of the photoacoustic response on soot concentration was possible. Therefore, the photoacoustic experiment provides information on the nature and quantity of soot produced by the combustion process.

A similar arrangement can be employed to monitor exhaust particulate from a diesel vehicle by photoacoustic spectroscopy [81]. In the experiments described in the following an argon laser with 1,5 W at 514,5 nm was the light source for the resonant photoacoustic experiment. A scheme of the experimental setup is shown in Fig. 25 [82]. The main constituents of diesel exhaust particulate material are graphitic carbon with various organic compounds adsorbed onto the carbon, about 1 percent trace metals, and about 2 percent sulfates. The fraction of organic material varies with engine type and operating conditions. Because only absorbed light produces a photoacoustic effect, this method essentially sees graphitic carbon. A linear dependence

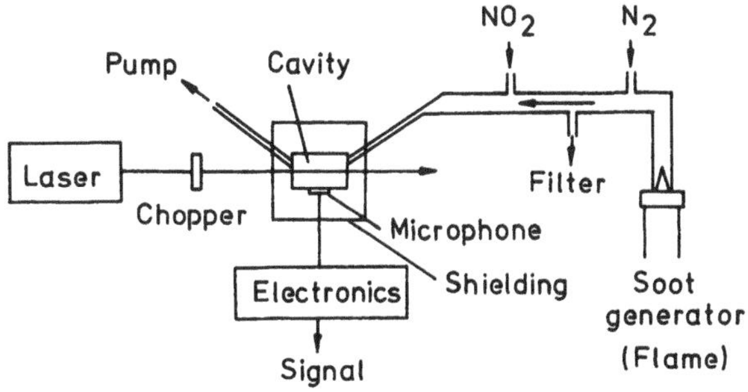

Fig. 24. Scheme of an experimental setup employed to detect photoacoustic spectra of airborne soot generated in a flame [79]. NO_2 was used as a reference; N_2 served as carrier gas; the total flow through the tube was 3,5 l/min; the airborne soot concentration was measured with a filter to be $\approx 1{,}4 \cdot 10^{-4}$ g/l

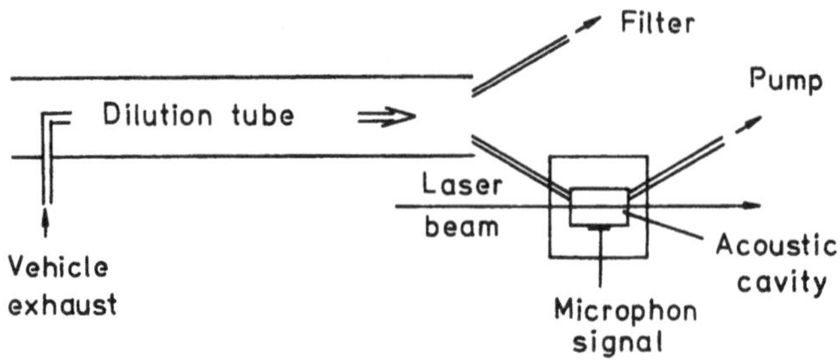

Fig. 25. Scheme of a dilution tube system and a resonant photoacoustic cell for the investigation of exhaust particulate [82]; the filter collection was used for chemical and mass analysis; the sampling rate for the photoacoustic analysis was ≈ 1,5 l/min

of the acoustic signal on laser intensity is observed if the microphone signal is plotted versus the average mass loading of graphitic particulate [81]. The mass-specific absorption coefficient for the graphitic particulate component of the diesel exhaust is calculated to be $8.9 \pm 0.5\ m^2/g$ at 514.5 nm. This value is in good agreement with the corresponding values derived for similar particulate material produced by other combustion sources. Roessler and Faxvog, for example, found an absorption coefficient of $8.3 \pm 0.9\ m^2/g$ by studying an acetylene flame with the spectrophone [80].

6 Summary

Nearly one hundred years after Bells discovery, resonant photoacoustic spectroscopy awoke from a long sleep. The introduction of lasers as radiation sources has led to rapid development of photoacoustics in the last ten years. Experimental results obtained under carefully optimized conditions are now in good agreement with theory of resonators. We can expect that new applications of resonant photoacoustic spectroscopy will be found in the near future. An important impetus in this research comes from the laser sources available. For many applications the doping procedure using a small amount of a highly absorbant gas in a mixture can be applied. This allows the manifold utilization of a single frequency laser. However, innovations in laser technology will open up new possibilities to photoacoustic research. In the past, the treatment of analytical problems such as trace gas analysis was dominant, but recently new applications of resonant photoacoustic spectroscopy have been reported and such progress can be expected to continue.

7 Acknowledgements

Support of our photoacoustic work by the Deutsche Forschungsgemeinschaft and the Fonds der Chemischen Industrie is gratefully acknowledged. The author also thanks Clifton Sayer for reading the manuscript.

8 References

1. Bell, A. G.: Am. J. Sci. *20*, 305 (1880)
2. Bell, A. G.: Phil. Mag. *11*, 510 (1881)
3. Tyndall, J.: Proc. Roy. Soc. London *31*, 307 (1881)

4. Röntgen, W. C.: Phil. Mag. *11*, 308 (1881)
5. Viengerov, M. L.: Dokl. Akad. Nauk. SSSR *19*, 687 (1938)
6. Luft, K. F.: Z. Tech. Phys. *24*, 97 (1943)
7. Gorelik, G.: Dokl. Akad. Nauk. SSSR *54*, 779 (1946)
8. Slobodskaya, P. V.: Izv. Akad. Nauk. SSSR Ser. Fiz. *12*, 656 (1948)
9. Kerr, E. L., Atwood, J. G.: Appl. Opt. *7*, 915 (1968)
10. Kreuzer, L. B.: J. Appl. Phys. *42*, 2934 (1971)
11. Dewey Jr., C. F., Kamm, R. D., Hackett, C. E.: Appl. Phys. Lett. *23*, 633 (1973)
12. Pao Yoh-Han: Optoacoustic spectroscopy and detection, New York, Academic Press 1977
13. Rosencwaig, A.: Photoacoustics and photoacoustic spectroscopy, New York, Wiley & Sons 1980
14. Lambert, J. D.: Vibrational and rotational relaxation in gases, Oxford, Clarendon Press 1977
15. Huetz-Aubert, M., Chevalier, P.: Adv. Mol. Relax. Proc. *2*, 101 (1970); Huetz-Aubert, M., Chevalier, P., Tripodi, R.: J. Chem. Phys. *54*, 2289 (1971)
16. Hess, P., Moore, C. B.: J. Chem. Phys. *65*, 2339 (1976)
17. Mandich, M. L., Flynn, G. W.: J. Chem. Phys. *73*, 1265 (1980)
18. Janiesch, W., Kadibelban, R., Hess, P.: Chem. Phys. Lett. *71*, 387 (1980)
19. Finzi, J. et al.: J. Chem. Phys. *67*, 4053 (1977)
20. Frank, K., Hess, P.: Chem. Phys. Lett. *68*, 540 (1979)
21. Röper, J., Hess, P.: Appl. Phys. Lett. *39*, 946 (1981)
22. Karbach, A., Röper, J., Hess, P.: to be published
23. Goldan, P. D., Goto, K.: J. Appl. Phys. *45*, 4350 (1974)
24. Thomas III, L. J., Kelly, M. J., Amer, N. M.: Appl. Phys. Lett. *32*, 736 (1978)
25. Olson, D. B., Mallard, W. G., Gardiner Jr., W. C.: Appl. Spectros. *32*, 489 (1978)
26. Gerlach, R., Amer, N. M.: Appl. Phys. Lett. *32*, 228 (1978)
27. Max, E., Rosengren, L.-G.: Opt. Commun. *11*, 422 (1974)
28. Johnson, R. H. et al.: Appl. Opt., to be published
29. Shtrikman, S., Slatkine, M.: Appl. Phys. Lett. *31*, 830 (1977)
30. Gerlach, R., Amer, N. M.: Appl. Phys. *23*, 319 (1980)
31. Koch, K. P., Lahmann, W.: Appl. Phys. Lett. *32*, 289 (1978)
32. Moldover, M. R., Waxman, M., Greenspan, M.: High Temp., High Pressure *11*, 75 (1979)
33. Mehl, J. B., Moldover, M. R.: J. Chem. Phys. *74*, 4062 (1981)
34. Murphy, J. C., Aamodt, L. C.: J. Appl. Phys. *48*, 3502 (1977)
35. Monohan Jr., E. M., Nolle, A. W.: J. Appl. Phys. *48*, 3519 (1977)
36. Quimby, R. S., Selzer, P. M., Yen, W. M.: Appl. Opt. *16*, 2630 (1977)
37. Fernelius, N. C.: Appl. Opt. *18*, 1784 (1979)
38. Bechthold, P. S., Campagna, M., Chatzipetros, J.: Opt. Commun. *36*, 369 (1981)
39. Busse, G., Herboeck, D.: Appl. Opt. *18*, 3959 (1979)
40. Nordhaus, O., Pelzl, J.: Appl. Phys. *25*, 221 (1981)
41. Cahen, D., Lerner, E. I., Auerbach, A.: Rev. Sci. Instrum. *49*, 1206 (1978)
42. Bechthold, P. S., Campagna, M., Schober, T.: Solid State Commun. *36*, 225 (1980)
43. Shaw, R. W.: Appl. Phys. Lett. *35*, 253 (1979)
44. Zahn, R.: J. Acoust. Soc. Am. *69*, 1200 (1981)
45. Sessler, G. M., West, J. E.: J. Acoust. Soc. Am. *53*, 1589 (1973)
46. Rayleigh, J. W. S.: Theory of sound, New York, Dover 1894
47. Morse, P. M.: Vibration and Sound, New York, McGraw Hill 1948
48. Nodov, E.: Appl. Opt. *17*, 1110 (1978)
49. Mehl, J. B., Moldover, M. R.: J. Chem. Phys. *77*, 455 (1982)
50. Younglove, B. A., McCarty, R. D.: J. Chem. Thermodynamics *12*, 1121 (1980)
51. Kamm, R. D.: J. Appl. Phys. *47*, 3550 (1976)
52. Moldover, M. R., Mehl, J. B.: to be published in "Proceedings of the Sec. Internat. Conf. on Precision Measurements and Fundamental Constants, June 1981, Nat. Bur. Stand., Gaithersburg, MD.
53. Morse, P. M., Ingard, K. U.: Theoretical Acoustics, New York, McGraw Hill 1968
54. Frank, K., Hess, P.: Ber. Bunsenges. Phys. Chem. *84*, 724 (1980)
55. Mehl, J. B., Moldover, M. R.: Eighth Symposium on Thermophysical Properties, June 1981, Nat. Bur. Stand., Gaithersburg, MD.
56. Leslie, D. H., Trusty, G. L.: Appl. Opt. *20*, 1941 (1981)
57. Patel, C. K. N., Kerl, R. J.: Appl. Phys. Lett. *30*, 578 (1977)

58. Röper, J., Hess, P.: to be published in Photoacoustics — Principles and Applications, Wiesbaden, Vieweg-Verlag
59. Gammon, B. E.: to be published
60. Herzberg, G.: Electronic Spectra of Polyatomic Molecules, New York, Van Nostrand Co, 1966, summarized by Chao, J. and Zwolinski, B. J.: J. Phys. Chem. Ref. Data *4*, 251 (1975)
61. Shimanouchi, T.: Tables of Molecular Vibrational Frequencies, Consolidated Vol. I, NSRDS-NBS 39, 1972
62. Douslin, D. R., Harrison, R. H.: J. Chem. Thermodynamics *8*, 301 (1976)
63. Waxman, M., Davis, H. A.: in Adv. Chem. Ser. 182, Equations of State in Engineering and Research, (Chao, K. C., Robinson, R. L. Jr. eds.), Am. Chem. Soc. Washington DC, 1979, p. 285
64. Kadibelban, R., Hess, P.: Appl. Opt. *21*, 61 (1982)
65. Bruce, C. W. et al.: Appl. Opt. *15*, 2970 (1976)
66. Avramides, E., Hunter, T. F.: Chem. Phys. *57*, 441 (1981)
67. Avramides, E., Hunter, T. F.: to be published
68. Zharov, V. P., Montanari, S. G.: Opt. Spectros. *51*, 66 (1981)
69. Kraft, H. G., Bevan, J. W.: Infrared Phys. *21*, 337 (1981)
70. German, K. R., Gornall, W. S.: J. Opt. Soc. Am. *71*, 1452 (1981)
71. Rank, D. H. et al.: J. Opt. Soc. Am. *50*, 421 (1960)
72. Smith, W. H., Gelfand, J.: J. Quant. Spectros. Radiat. Transfer *24*, 15 (1980)
73. McKellar, A. R. W., Goetz, W., Ramsay, D. A.: Astrophys. J. *207*, 663 (1976)
74. Reddy, K. V., Heller, D. F., Berry, M. J.: J. Chem. Phys. *76*, 2814 (1982)
75. Reddy, K. V., Berry, M. J.: Second Internat. Topical Meeting on Photoacoustic Spectroscopy in Berkeley 1981, Technical Digest, paper MB4-1.
76. Chan, C. H.: Appl. Phys. Lett. *26*, 628 (1975)
77. Terhune, R. W., Anderson, J. E.: Opt. Lett. *1*, 70 (1977)
78. Truex, T. J., Anderson, J. E.: Atm. Env. *13*, 507 (1979)
79. Japar, S. M., Killinger, D. K.: Chem. Phys. Lett. *66*, 207 (1979)
80. Roessler, D. M., Faxvog, F. R.: J. Opt. Soc. Am. *69*, 1699 (1979)
81. Japar, S. M., Szkarlat, A. C.: Combust. Sci. Technol. *24*, 215 (1981)
82. Japar, S. M., Szkarlat, A. C.: Fuels & lubricants meeting Tulsa, Oklahoma 1981, SAE technical paper series

Non-Aqueous Electrolyte Solutions in Chemistry and Modern Technology

Josef Barthel, Heiner-J. Gores, Georg Schmeer, and Rudolf Wachter

Institut für Physikalische und Theoretische Chemie der Universität Regensburg, FRG

Table of Contents

Part A
Fundamentals of Chemistry and Physical Chemistry of Non-Aqueous Electrolyte Solutions

Part B
Technical Applications and Applied Research

In this paper a brief survey is given of the properties of non-aqueous electrolyte solutions and their applications in chemistry and technology without going into the details of theory. Specific solvent-solute interactions and the role of the solvent beyond its function as a homogenous isotropic medium are stressed. Taking into account Parker's statement [1] "*Scientists nowadays are under increasing pressure to consider the relevance of their research, and rightly so*" we have included examples showing the increasing industrial interest in non-aqueous electrolyte solutions.

The concepts and results are arranged in two parts. Part A concerns the fundamentals of thermodynamics, transport processes, spectroscopy and chemical kinetics of non-aqueous solutions and some applications in these fields. Part B describes their use in various technologies such as high-energy batteries, non-emissive electro-optic displays, photoelectrochemical cells, electrodeposition, electrolytic capacitors, electro-organic synthesis, metallurgic processes and others.

Four Appendices are added. Appendix A gives a survey on the most important non-aqueous solvents, their physical properties and correlation parameters, and the commonly used abbreviations. Appendices B and C show the mathematical background of the general chemical model. The symbols and abbreviations of the text are listed and explained in Appendix D.

Part A

Fundamentals of Chemistry and Physical Chemistry of Non-Aqueous Electrolyte Solutions

I Introduction

The systematic investigation of non-aqueous solutions is guided by the progress of our knowledge on solute-solute and solute-solvent interactions. By combination with chemical models of the solution, valuable results can be obtained which assist the understanding of the properties of these solutions.

For *Dilute Electrolyte Solutions* consistent and reliable equations are based on the modern conception of electrochemistry which takes into account both long and short-range forces between the solute and solvent particles in the framework of the McMillan-Mayer-Friedman approaches to theory [2]. Solution chemists usually think of short-range interactions in terms of ion-pair formation. A chemical model of electrolyte solutions as developed and used in our laboratory [3,4] is the basis of the fundamentals given in part A of this survey. It allows the use of the classical association concept initially introduced by Bjerrum [5] after some refinements concerning the spatial extension and structure of ion pairs and the mean-force potentials. Classical thermodynamics and transport phenomena are unable to distinguish between ion pairs and undissociated electrolyte molecules, both proving their presence in the solutions as neutral particles in equilibrium with the 'free' ions. However, in favourable cases ion pairs may be detected separately from undissociated electrolyte molecules by spectroscopic methods. The ions in an ion pair retain their individual ionic characters and are linked only by Coulombic and short-range forces, including H-bonding.

Modern developments of the statistical-mechanical theory of solutions provide valuable results, but no satisfactory answer can yet be found to fundamental questions such as the effect of ions on the permittivity of the solvent or on the structures in solution [2]. Computer simulations may be helpful in understanding how some fundamental properties of the solutions are derived from fundamental laws. However, the actual limitation to a set of a few hundred particles in a box of about 20 Å of length, a time scale of the order of picoseconds, and pair potentials based on classical mechanics usually prevent the deduction of useful relationships for the properties of real electrolyte solutions.

The treatment of *Electrolyte Solutions from Moderate to Saturated Concentrations* either rationalizes the effects from ion-ion and ion-solvent interactions in terms of the parameters characterizing the behaviour of electrolytes in dilute solutions or uses correlation methods based on empirical interaction scales [6-21]. Examples of both procedures will be given.

II Classification of Solvents and Electrolytes

Any attempt to set up a complete theory which takes into account all types of interaction energy between the ions and molecules of an electrolyte solution with the aim of determining the properties of the solution by means of statistical thermodynamic methods must be unsuccessful as a result of insuperable mathematical difficulties. The approaches to the problem which have been used were outlined in Section I.

Two aspects determine the role of the solvent: its bulk properties and its electron donor or acceptor abilities. The Debye-Hückel theory [22] which is valid at infinitely low concentrations, recognizes solvents only by their bulk properties, *i.e.* relative permittivity ε, viscosity η, and density ϱ. However, the Debye-Hückel range of validity is often experimentally unattainable (Ref. [4], cf. also Figs. 4 and 6). The importance of bulk properties decreases with increasing electrolyte concentration.

Donor and acceptor properties are the main factors which govern processes on the molecular scale, *i.e.* solvation and association. Theoretical and semiphenomenological approaches use molecular properties, dipole and quadrupole moments, polarizability etc., or mean-force potentials for taking these effects into consideration. Applied solution chemistry takes account of them with the help of the empirical scales previously mentioned [6–21].

Various attempts have been made to classify solvents, e.g. according to bulk and molecular properties [20], empirical solvent parameter scales [6–21], hydrogen-bonding ability [23,24], and miscibility [25]. In table I solvents are divided into classes on the basis of their acid-base properties [26–29] which can be used as a general chemical measure of their ability to interact with other species. Detailed information on these and other solvents, their symbols, fusion and boiling points (θ_F and θ_B), bulk properties (ε, η, ϱ), and currently-used correlation parameters DN (donor number), E_T-value, and AN (acceptor number) is given in Appendix A-1.

Table I Classification of organic solvents (for detailed information see: Appendix A-1)

Solvent class	Examples
1. amphiprotic hydroxylic	methanol (MeOH); ethanol (EtOH); 1-propanol (PrOH); diethylene glycol (DEG); glycerol;
2. amphiprotic protogenic	acetic acid
3. protophilic H-bond donor	formamide (FA); N-methylformamide (NMF); diaminoethane;
4. aprotic protophilic	dimethylformamide (DMF); 1-methyl-2-pyrolidone (NMP); hexamethylphosphoric triamide (HMPT); dimethylsulfoxide (DMSO)
5. aprotic protophobic	acetonitrile (AN); sulfolane (TMS); propylene carbonate (PC); γ-butyrolactone (γ-BL); acetic anhydride
6. low permittivity — electron donor	diethyl ether (DEE); tetrahydrofurane (THF); diglyme (DG); 1,2-dimethoxyethane (DME); 1,4-dioxane
7. inert	dichloromethane; tetrachloroethylene, benzene, cyclohexane

The symbols quoted there, e.g. MeOH, THF, DME etc., are used in this text. No classification is universally applicable. Overlapping of the solvent classes is inevitable and some specific solute-solvent interactions evade classification. Specific interactions, however, are often sought in connexion with technological problems and have led to a search for appropriate solvent mixtures which are gaining importance in many fields of applied research. In spite of all its limitations, the classification of solvents is useful for rationalizing the choice of appropriate solvents and solvent mixtures for particular investigations.

Electrolytes can be classified in two categories, ionophores [30] (true electrolytes [31]) and ionogenes [30] (potential electrolytes [31]). Ionophores are substances which, in the pure state, already exist as ionic crystals, e.g. alkali metal halides. Ionogenes, such as carboxylic acids, form ions only by chemical reactions with solvent molecules. Amphiprotic solvents themselves behave as ionogenes in producing their lyonium ions (anions) and lyate ions (cations) by autoprotolysis reactions. Ionophores are initially completely dissociated in solution and their ions are solvated. However, almost all solvents allow ion-association to ion pairs and higher ion aggregates, both with and without inclusion of solvent molecules, to occur. When electrically neutral these species cannot transport current. The following examples are given for illustration:

(a) ionophores:

$$\mathrm{NaCl(cryst)} \xrightarrow[\text{acetonitrile}]{} \mathrm{Na^+ + Cl^-} \overset{K_A}{\rightleftharpoons} [\mathrm{Na^+Cl^-}]^0$$

$$\mathrm{LiBF_4(cryst)} \xrightarrow[\text{dimethoxyethane}]{} \mathrm{Li^+ + BF_4^-} \overset{K_A}{\rightleftharpoons} [\mathrm{Li^+BF_4^-}]^0$$

and further association to $[Li^+(BF_4^-)_2]^-$ etc.

(b) ionogenes:

$$\mathrm{HClO_4 + CH_3COOH} \overset{K_I}{\rightleftharpoons} [\mathrm{CH_3COOH_2^+ClO_4^-}]^0 \overset{K_D}{\rightleftharpoons} \mathrm{CH_3COOH_2^+ + ClO_4^-}$$

$$\mathrm{CH_3OH + CH_3OH} \overset{K_S}{\rightleftharpoons} \mathrm{CH_3OH_2^+ + CH_3O^-}.$$

These association and dissociation reactions do not usually proceed to completion. Both processes are described by the thermodynamic equilibrium constants K_A (association constant) or K_D (dissociation constant). The dissolution of perchloric acid in glacial acetic acid [32–34] shows the typical ionisation equilibrium (equilibrium constant, K_I) preceding the dissociation process in the case of ionogenes. The overall constant K is given by

$$K = \frac{K_I K_D}{1 + K_I}. \qquad (1)$$

For strong acids and bases, where $K_I \gg 1$, equation (1) reduces to $K = K_D$, whereas for weak ones, where $K_I \ll 1$, $K = K_I K_D$. The sequence of basicities can change with changing solvent; the K_I-values are more significant for the discussion of ionogenes than are the K-values. Ionisation constants, however, will not be discussed in this article. It is sufficient to note that the dissolution of acids or bases in amphiprotic solvents is generally followed by a protolysis reaction (partial neutralisation) as a consequence of the following equilibria:

$$AH + SH \rightleftarrows A^- + SH_2^+ \tag{2a}$$

$$B + SH \rightleftarrows BH^+ + S^- \tag{2b}$$

where SH represents the amphiprotic solvent, AH the acid and B the base. When these equilibria are shifted markedly towards ionisation, a levelling effect [35] occurs which almost completely replaces the acids or bases with the lyate or lyonium ions of the solvent, irrespective of the initial strong acid or base. A typical example of this is the behaviour of the mineral acids in water.

The series of simultaneous equilibria including K_I, K_D (or K_A) and the autoprotolysis constant K_S limits the quantitative discussion of electrolyte solutions to simple cases. However, the appropriate choice of such cases will give valuable insight into the properties of electrolyte solutions, especially those of ionophores where the ionisation step need not be considered.

Finally, a classification of the individual ions is only possible in a rough and incomplete way. Monoatomic cations can be arranged according to the number of electrons and are referred to as "d^n-cations" (e.g. alkali metal cations then are d^0-cations). Cations such as $[R_4N]^+$, $[R_4P]^+$, or $[R_3S]^+$ have their charges shielded by alkyl or aryl groups. They are almost non-polarizable and are referred to as inert cations. Cations of type $[R_{4-n}H_nN]^+$ or $[(ROH)_4N]^+$ are protic cations, capable of forming H—bonds with anions or solvent molecules. The distinction between "hard" and "soft" cations has no significant relevance here, but for anions it is useful. A detailed discussion of the properties of electrolyte solutions as a function of ion classes and solvents is given in Ref. [36].

It should be mentioned that experimental investigations, especially in dilute non-aqueous solutions, require highly purified solvents and solutes. Impurities can change the properties of the solution drastically. A water content of 20 ppm is equivalent to the total amount of solute in a 10^{-3} molar solution. For checking the purity of solvents UV cut-off, conductivity, chromatography as well as thermal and electrochemical methods are recommended [37–41]. The control of the purity of electrolytes is more difficult; for details see Refs. [37–42].

III Thermodynamics of Electrolyte Solutions

The thermodynamic properties of an electrolyte solution can be derived from the chemical potentials μ_i of its components which are given by the relationships [43], cf. also [36]

$$\mu_s(p, T) = \mu_s^*(p, T) + RT \ln x_s f_s ; \tag{3a}$$

$$\mu_s^*(p, T) = \lim_{x_s \to 1} \mu_s(p, T); \qquad \lim_{x_s \to 1} f_s = 1 \tag{3b, c}$$

for the solvent S, and

$$\mu_i(p, T) = \mu_i^{\infty(c)}(p, T) + \nu_i\, RT \ln c_\pm^{(i)} y_\pm^{(i)}\,; \qquad i = 1, 2, \ldots \tag{4a}$$

$$\mu_i^{\infty(c)}(p, T) = \lim_{x_s \to 1} [\mu_i(p, T) - \nu_i\, RT \ln c_\pm^{(i)}]; \qquad \lim_{x_s \to 1} y_\pm^{(i)} = 1 \tag{4b, c}$$

for the electrolyte components $Y_i = (X_1^{z_1})_{\nu_1^i} (X_2^{z_2})_{\nu_2^i} (X_3^{z_3})_{\nu_3^i} \ldots$ which dissociate into their ions $X_m^{z_m}$ according to the scheme $Y_i \to \sum_m \nu_m^i X_m^{z_m}$.

In these equations, x_s and $c_\pm^{(i)}$ are the mole fraction of the solvent, S, and the mean concentrations [mol/dm^3 of solution] of the electrolytes, Y_i; f_s and $y_\pm^{(i)}$ are the appropriate activity coefficients

$$c_\pm^{(i)} = \prod_m c_m^{\nu_m^i/\nu_i}; \qquad y_\pm^{(i)} = \prod_m y_m^{\nu_m^i/\nu_i}; \qquad \nu_i = \sum_m \nu_m^i\,. \tag{4d, e, f}$$

Eqs. (4a) are replaced by an equivalent set of equations

$$\mu_i(p, T) = \mu_i^{\infty(m)} + \nu_i\, RT \ln m_\pm^{(i)} \gamma_\pm^{(i)}\,; \qquad i = 1, 2, \ldots \tag{5}$$

if mean molalities, $m_\pm^{(i)}$ [mol/kg of solvent] are chosen as the concentration scale with $\gamma_\pm^{(i)}$ as the appropriate activity coefficients. Eqs. (4b–f) then must be appropriately changed. Conversion formulae for the reference chemical potentials, concentration scales and activity coefficients are given in Refs. [31,36,43,44].

The relationships 3 and 4, or 3 and 5, respectively, form a complete set of equations on which a general discussion of the thermodynamic properties of the electrolyte solution can be based.

As only some fundamental ideas can be illustrated in this article, the following discussion is restricted to solutions of a solvent S and a symmetrical electrolyte $Y = C^{z+}A^{z-}$ yielding z_+-valent cations C^{z+} and z_--valent anions A^{z-}. The chemical potential of the solute is then given by the relationships

$$\mu_Y(p, T) = \mu_Y^{\infty(c)}(p, T) + 2\, RT \ln c_\pm y_\pm \tag{6a}$$

$$c_\pm = [c_+ c_-]^{1/2} = c_Y = c\,; \qquad y_\pm = [y_+ y_-]^{1/2} \tag{6b, c}$$

or by a similar set of equations derived from eq. (5).

Eq. (6a) is valid for both completely and partially dissociated (or associated) electrolytes if the activity coefficient is written as follows

$$y_\pm = \alpha y'_\pm \tag{7}$$

with α as the degree of dissociation and $y'_\pm$ as the activity coefficient of the dissociated part (free ions) of the electrolyte component of the solution [4,36,44]. For completely dissociated electrolytes α is equal to 1. The ion pairs of ionophoric electrolytes and the molecules of ionogenic electrolytes are in equilibrium with the free ions. The equilibrium condition yields the equilibrium constant as an association (K_A) or dissociation (K_D) constant

$$K_A = K_D^{-1} = \frac{1 - \alpha}{\alpha^2 c y_{\pm}'^2} \tag{8}$$

The solution process of n_Y mol of electrolyte Y in n_s mol of solvent S is accompanied by a change in Gibbs energy

$$\Delta_{sol}G = n_s(\mu_s - \mu_s^*) + n_Y(\mu_Y - \mu_Y^*) \, . \tag{9}$$

Analogous equations are obtained for every other extensive thermodynamic property Z, *i.e.*

$$\Delta_{sol}Z = n_s(Z_s - Z_s^*) + n_Y(Z_Y - Z_Y^*) \tag{10}$$

where Z_i and Z_i^* are the partial molar quantities of the solvent or the electrolyte in the solution (Z_i) and in the pure phase (Z_i^*). Another useful formulation of $\Delta_{sol}Z$ is based on the definition of apparent molar quantities Φ_Z of the solute

$$Z = n_s Z_s^* + n_Y \Phi_Z \tag{11}$$

yielding the relationship

$$\frac{\Delta_{sol}Z}{n_Y} = \Delta_{sol}Z_Y^{\infty} + (\Phi_Z - \Phi_Z^{\infty}); \qquad \Delta_{sol}Z_Y^{\infty} = Z_Y^{\infty} - Z_Y^* \tag{12a, b}$$

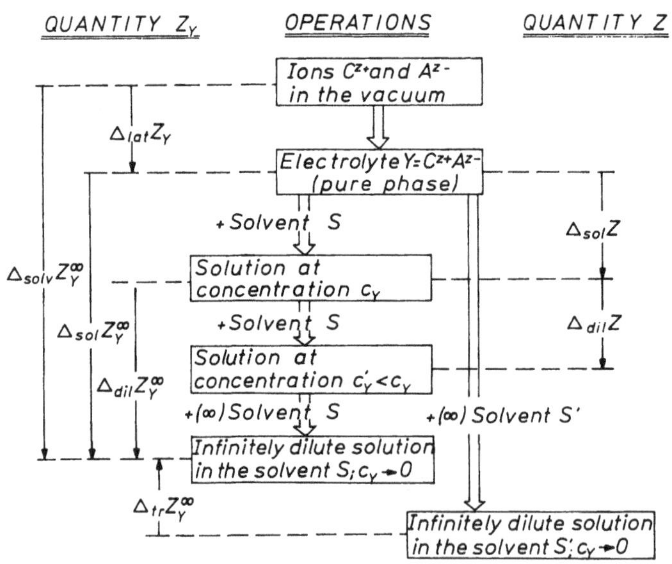

Fig. 1. Operations on electrolyte solutions at constant pressure and temperature and their appropriate translation into thermodynamic quantities:

$$\Delta Z = Z_{fin.} - Z_{init.}$$

lat: lattice; solv: solvation; sol: solution; dil: dilution; tr; transfer from solvent S' to solvent S. Z = V (volume); S (entropy); C_p (heat capacity); H (enthalpy); G (Gibbs energy) if the thermodynamic relation $\Phi_Z^\infty = Z_Y^\infty$ is taken into account. The molar quantity $\Delta_{sol}Z_Y^\infty$ corresponds to the transfer of 1 mol electrolyte compound from its pure state to infinite dilution in the solvent S.

Figure 1 summarizes the operations and notions used in solution chemistry.

IV Short and Long-Range Forces in Dilute Electrolyte Solutions

4.1 Distribution Functions and Mean-Force Potentials

The statistical theory of electrolyte solutions is built up around the distribution functions of the ions, cf. [30,44–49]. Electrolytes in solution give ions of the types $X_1^{z_1}$, $X_2^{z_2}$, ... in the analytical concentrations N_1, N_2, ... ions/cm^3. The distribution of the ions in the solution depends on the forces acting between all the particles, ions and solvent molecules. External forces are also involved in the description of transport properties.

The analytical concentrations, N_i, are one-particle molecular distribution functions and do not provide any information on particle interactions. Two-particle molecular distribution functions

$$f_{ij}(\vec{r}_1, \vec{r}_2) = N_i N_j g_{ij}(\vec{r}_1, \vec{r}_2) = N_j N_i g_{ji}(\vec{r}_2, \vec{r}_1) = f_{ji}(\vec{r}_2, \vec{r}_1) \tag{13}$$

indicate the probability of finding two ions, X_i and X_j, simultaneously at points $P_1(\vec{r}_1)$ and $P_2(\vec{r}_2)$ in the solution, regardless of the position of the remaining ions and regardless of the velocities of all the particles, see Fig. 2. The pair-correlation

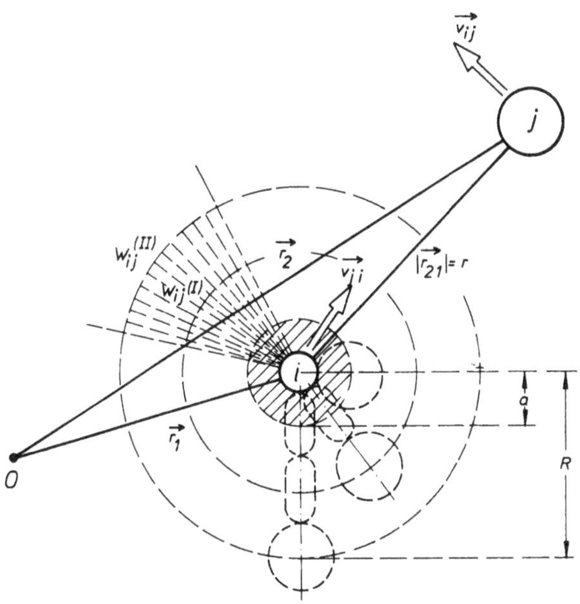

Fig. 2. The chemical model of electrolyte solutions. O: observer, i: ion X_i; j: ion X_j in an arbitrary position, $\vec{r}_{21}$, with regard to the ion X_i; special positions (contact, separation by one or two orientated solvent molecules) are sketched with broken lines. r, a, R: distance parameters; W_{ij}: mean-force potentials; $\vec{v}_{ij}$ and $\vec{v}_{ji}$: relative velocities of ions X_i and X_j

functions, $g_{ij}(\vec{r}_1, \vec{r}_2)$, are related to the mean-force potentials $W_{ij}(\vec{r}_1, \vec{r}_2)$ between the ions X_i and X_j by the relationship

$$W_{ij}(\vec{r}_1, \vec{r}_2) = -kT \ln g_{ij}(\vec{r}_1, \vec{r}_2) . \tag{14}$$

In addition they can be used to determine the local concentrations N_{ij} of ions X_j at a distance $\vec{r}_{21}$ ($\vec{r}_{21} = \vec{r}_2 - \vec{r}_1$) from an ion X_1 situated at $\vec{r}_1$

$$N_{ij}(\vec{r}_1, \vec{r}_{21}) = N_j g_{ij}(\vec{r}_1, \vec{r}_{21}) . \tag{15}$$

The relative velocities (see Fig. 2) of ions X_i and X_j, $\vec{v}_{ij}(\vec{r}_1, \vec{r}_{21})$ and $\vec{v}_{ji}(\vec{r}_2, \vec{r}_{12})$, and the two-particle molecular distribution functions are linked by Onsager's continuity equation [45]:

$$-\frac{\partial f_{ij}}{\partial t} = \operatorname{div}_1 [f_{ij}\vec{v}_{ij}] + \operatorname{div}_2 [f_{ji}\vec{v}_{ji}] = -\frac{\partial f_{ji}}{\partial t} \tag{16}$$

where the differential operators are applied with regard to the coordinates in $\vec{r}_1$ and $\vec{r}_2$. A general treatment yielding Eq. (16) starts from Liouville's theorem and uses the BBGKY hierarchy of equations [48,52].

For basic information on electrolyte theory see Refs. [2-4,22,30,36,44-70].

4.2 The Basic Chemical Model of Electrolyte Solutions

In order to obtain a framework which allows the development of appropriate equations for the properties of solutions at low electrolyte concentrations, a model of the ions and their surroundings must be used which takes into account both short and long-range forces. For this purpose the space around an ion is subdivided into three regions (see Fig. 2).

i) $r \leqq a$, a being the minimum distance of two oppositely-charged ions which is assumed to be the sum of effective cation and anion radii, $a = a_+ + a_-$.
ii) $a \leqq r \leqq R$, within which a paired state of oppositely-charged ions, the so-called ion pair, suppresses long-range interactions with other ions in the solution. In dilute solutions the occupation of the region $a \leqq r \leqq R$ by ions of the same sign or by more than two ions can usually be neglected.
iii) $r \geqq R$, the region of long-range ion-ion Coulombic interactions.

Table II shows the mean-force potentials for a dilute solution of the electrolyte compound $Y = C^{z+}A^{z-}$.

The model is a McMillan-Mayer (MM)-level Hamiltonian model. Friedman [2] characterizes models of this type as follows: "With MM-models it is interesting to see whether one can get a model that economically and elegantly agrees with all of the relevant experimental data for a given system; success would mean that we can understand all of the observations in terms of solvent-averaged forces between the ions. However, it must be noted that there is no reason to expect the MM potential function to be nearly pairwise additive. There is an upper bound on the ion concentration range within which it is sensible to compare the model with data for real systems if the pairwise addition approximation is made."

Table II Mean-force potential $W_{ij}(r)$ of ion—ion interaction in dilute solutions of symmetrical electrolytes

The mean-force potentials of this table are special formulae of the general charge distribution, Appendix B, for a single charge (ion).

$$\varkappa^2 \text{ (SI-units)} = \frac{1000\, N_A e_0^2}{\varepsilon_0 \varepsilon k T} \times \Gamma; \quad \Gamma \text{ (ional concentration)} = \sum_j (\alpha c_j)\, z_j^2$$

ε_0: permittivity of vacuum; k: Boltzmann constant; e_0: charge of proton; N_A: Avogadro number; W^*_{ij} mean-force potential of the short-range forces.

Region	Mean-force potential
$r \leqq a$	∞
$a \leqq r \leqq R$	$\frac{e_0^2 z_i z_j}{4\pi\, \varepsilon_0 \varepsilon} \times \frac{1}{r} - \frac{e_0^2 z_i z_j}{4\pi\, \varepsilon_0 \varepsilon} \frac{\varkappa}{1 + \varkappa R} + W^*_{ij}$
$r \geqq R$	$\frac{e_0^2 z_i z_j}{4\pi\, \varepsilon_0 \varepsilon} \times \frac{1}{r} \times \frac{\exp[\varkappa(R - r)]}{1 + \varkappa R}$

A multitude of MM-level Hamiltonian models can be found for the same system. The features of our chemical model are given in Refs. [3,4,72].

i) The distance parameters a (minimum distance of two ions) and R (upper limit of the structured region around an ion) are fixed by chemical evidence.

iα) The lower distance is fixed as a hard-core radius by the center-to-center distance of the ions where these exist (e.g. alkali halides) or else is calculated from bond lengths or van der Waals volumes (e.g. tetraalkylammonium salts). For unsymmetrical ions like $Me_2Bu_2N^+$ or $C_2H_5O^-$ the shortest possible distance is taken to be the distance of closest approach. Some ions, e.g. Li^+ in water or protic solvents, require the inclusion of a functional group of the solvent molecule (here: OH) into the distance of closest approach. This leads to structures like $Li^+(ROH)\, Cl^-$ with $a = a_{cryst} + d_{OH}$.

iβ) The upper distance R is obtained by adding the length of one or more orientated solvent molecules to the distance of closest approach: $R = a + ns$, $n = 1, 2, \ldots$ Values a and s are quoted in Appendix A-2.

ii) The mean-force potentials of W_{ij} (Eq. 14) are split into two parts representing Coulombic ion-ion interaction, W^{el}_{ij}, and short-range interactions, W^*_{ij}. A further subdivision of the W^*_{ij}'s which specifies contributions from induction, dispersion and chemical forces (e.g. H—bonding) is possible. At the current stage of investigation, the contributions W^{el}_{ij} for every region are obtained from the resolution of a set of Poisson-equations and appropriate boundary conditions and the W^*_{ij} are chosen as step potentials.

iii) Extensions

iiiα) A subdivision of the region $a \leqq r \leqq R$ is useful when more than one shell of solvent molecules is orientated.

iiiβ) The introduction of local permittivities is possible.

iiiγ) The spherical charge symmetry can be replaced by an arbitrary charge distribution leading to angular-dependent potentials. Appendix B gives a summary of the appropriate potentials. Chemical kinetics makes use of this type of extension when kinetic salt, solvent, and substituent effects are treated for reactions between particles with complex charge distributions (see Sect. VIII).

4.3 The Ion-Pair Concept

Models of the electrolyte solution allow the introduction of the association concept if a critical distance around the central ion can be defined within which pair configurations of oppositely charged ions are considered as ion pairs. The link between the model and the experimentally determined thermodynamic property of the solution is an integral expression which can be subdivided in various ways

$$\int_0^\infty r^2 g_{ij}\, dr = \int_a^R r^2 \exp\left[-\frac{W_{ij}^{(I)}}{kT}\right] dr + \int_R^\infty r^2 \exp\left[-\frac{W_{ij}^{(II)}}{kT}\right] dr. \qquad (17)$$

The choice of R is arbitrary within reasonable limits and then divides up the thermodynamic excess function, $\mu_Y^E = \nu RT \ln y_\pm$, into contributions from the so-defined ion pair (degree of dissociation, α) and from the 'free' ions (activity coefficient of the free ions, $y'_\pm$), cf. Eq. (7). Onsager characterized the situation as follows [73]: "The distinction between free ions and associated pairs depends on an arbitrary convention. Bjerrum's choice is good, but we could vary it within reason. In a complete theory this would not matter; what we remove from one page of the ledger would be entered elsewhere with the same effect."

Theory alone cannot provide a criterion for the best association constant. However, the variety of solution models leads usually to more or less satisfactory association constants when all of the relevant experimental data including their dependence on temperature and pressure are considered. Once more, chemical evidence is a good criterion for the selection of the appropriate model. It will be shown in the following sections that the identification of the "critical distance of association" with the cut-off distance of the short-range forces, R, in our chemical model yields association constants which are almost independent of the experimental method of their determination.

The association concept is based on the equilibrium of 'free' ions and ion pairs in the solution

$$C^{z+} + A^{z-} \rightleftarrows [C^{z+}A^{z-}]^0 . \qquad (18)$$

The concentration equilibrium constant of ion-pair formation, $K_c = K_A y'^2_\pm$ (see Eq. (8)), can be written [4] as shown in the first equation of Fig. 3 when using the reduced partition functions Q_p, Q_+, and Q_- of ion pairs, free cations and free anions.

As a first approximation the free cations and anions are considered as charged spheres of masses m_+ and m_-. The ion pair is represented by an uncharged

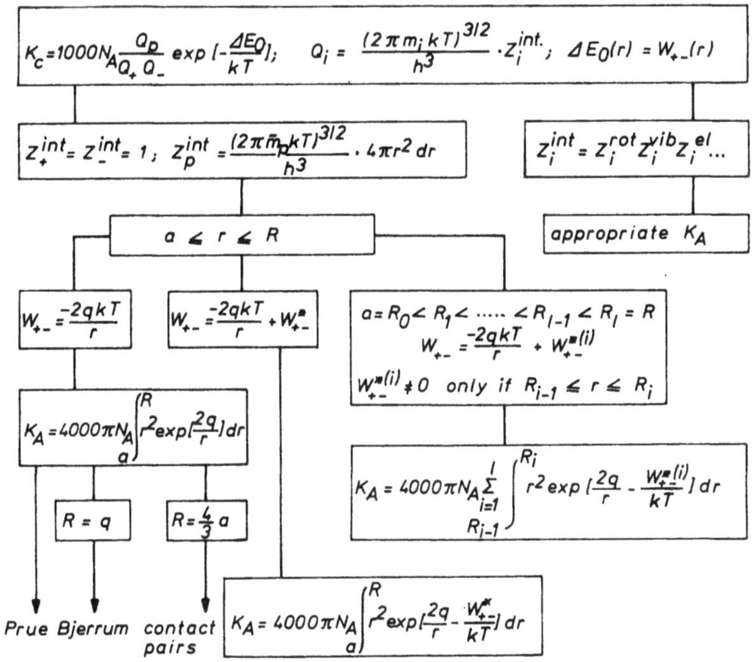

Fig. 3. The family tree of association constants

particle of reduced mass, $\tilde{m}_p$ ($\tilde{m}_p^{-1} = m_+^{-1} + m_-^{-1}$), in a spherical box of radius R, meaning that an ion pair is formed if two oppositely-charged ions have approached to within a distance smaller than R. The difference in energies of the reacting species, ΔE_0, can be identified with the mean-force potential W_{+-} ($a \leqq r \leqq R$), Table II.

Finally, cf. Fig. 3, the equation

$$K_A = 4000\pi N_A \int_a^R r^2 \exp\left[\frac{2q}{r} - \frac{W_{+-}^*}{kT}\right] dr; \qquad q = \frac{e_0^2 |z_+ z_-|}{8\pi\varepsilon_0 \varepsilon kT} \qquad (19a, b)$$

is obtained by this approximation. The activity coefficient of the dissociated part of the electrolyte, $y'_\pm$, is given by the relationship [3, 65]

$$y'_\pm = \exp\left[-\frac{\varkappa q}{1 + \varkappa R}\right]. \qquad (19c)$$

Figure 3 shows the family tree of some association constants which can be found in the literature and indicates the presuppositions for deducing them from the initial equation. For example, Bjerrum's association constant [5] and its appropriate activity coefficient are obtained from Eqs. (19) by setting $R = q$ and $W_{+-}^* = 0$. As a further example, the subdivision of the region $a \leqq r \leqq R$ into $a < R_1 < \ldots < R_{n-1} < R$ is straightforward [4]. A current application is the case of stepwise ion-pair formation [74-76], e.g.

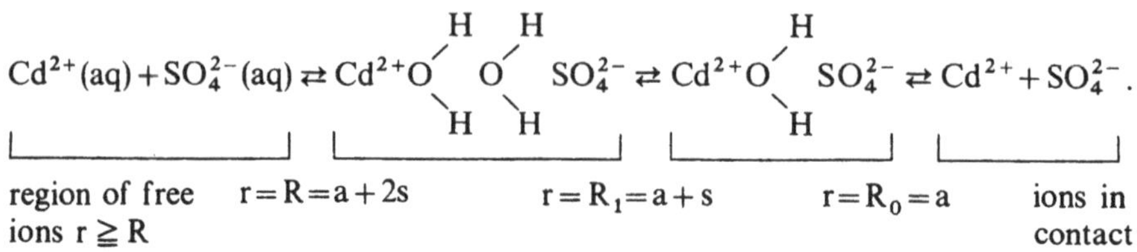

(20)

The distance parameter $a = a_+ + a_-$ has its usual meaning, s is the length of an orientated solvent molecule.

Association constants, as determined by thermodynamic or transport process mesurements, are the basis for determining the short-range forces around the ions and mean-force potentials in dilute solutions. These experiments provide distance parameters, R, as well as K_A-values and thus permit determination of short-range interaction potentials, W^*_{+-}, via Eq. (19a), or an equivalent expression from Fig. 3 [3,4,72,74-81]. On the other hand, molar quantities $\Delta G^*_A = N_A W^*_{+-}$ can be interpreted as the non-Coulombic part of the Gibbs energy of the ion-pair formation reaction [3,4,74,75]. In addition, calorimetric measurements and studies of the temperature-dependence of conductance provide enthalpies and entropies of ion-pair formation which yield valuable information. For example, alkali metal, alkaline earth and other divalent cation salts in protic solvents yield positive ΔH^*_A and highly positive ΔS^*_A values in contrast to tetraalkylammonium salts which show $\Delta H^*_A < 0$ and small entropies, ΔS^*_A [4,74-77]. Ion-pair formation within the former groups of salts involves the rearrangement of the solvation shells whereas that of the latter one scarcely does at all. For more details on comprehensive series of measurements see Refs. [4,77].

V Thermodynamic Properties of Electrolyte Solutions

5.1 Generalities

Properties E (c; p,T) of dilute and moderately concentrated electrolyte solutions (concentration range $\varkappa q < 0.5$; $\varkappa$ (Table II), q (Eq. 19b)), e.g. thermodynamic properties Z, can be represented by a set of equations [72]

$$E(c; p, T) = E^{\infty}(p, T) + E'(\alpha c, R; p, T) \tag{21a}$$

$$K_A = \frac{1-\alpha}{\alpha^2 c}\,\frac{1}{y'^2_{\pm}}\,; \qquad y'_{\pm} = \exp\left[-\frac{q\varkappa}{1+\varkappa R}\right] \tag{21b, c}$$

where $E^{\infty}(p, T)$ is the corresponding property of the infinitely dilute solution. The other symbols have the meaning given in the preceding sections. Transport properties are controlled by a similar set of equations, see Section VI.

The chemical model allows the determination of both values of $E^\infty(p, T)$ by a well-founded extrapolation method and values of R and W^*_{ij} independently of the special thermodynamic or transport property which is being investigated and thus provides data for other ones [3,72,77,81-83]. This fact can be used for computer methods which require the storing of a minimum of basic data to make these properties available. The data bank ELDAR [83] works on this principle. Spectroscopic or kinetic investigations do not necessarily furnish the complete set of parameters for establishing function E′, Eq. (21a), see Sections VII and VIII.

5.2 Solution and Dilution Experiments

The basic equations of type Eq. (21a) needed for solution and dilution experiments are obtained from Eqs. (12a, b). The apparent molar quantity Φ_Z may be split into two parts, one for the 'free' ions of the chemical model, $\Phi_Z(FI) = \Phi_Z(r > R)$, the other one for the ion pairs, $\Phi_Z(IP) = \Phi_Z(a \leqq r \leqq R)$. This assumption leads to the relationship

$$\Phi_Z - \Phi_Z^\infty = \Phi_Z^{rel} = [\alpha\Phi_Z(FI) + (1 - \alpha)\,\Phi_Z(IP)] - \Phi_Z^\infty \tag{22}$$

which is introduced into Eq. (12a) yielding the equation

$$\frac{\Delta_{sol}Z}{n_Y} = \Delta_{sol}Z_Y^\infty + \alpha\Phi_Z^{rel}(FI) + (1 - \alpha)\,\Delta Z_A^0\,. \tag{23}$$

Φ_Z^{rel} is the corresponding relative apparent molar quantity. For symmetrical electrolytes at moderate and low concentrations the quantity $\Phi_Z^{rel}(IP)$ equals $\Delta Z_A^0 = \Phi_Z(IP) - \Phi_Z^\infty$, *i.e.* the molar quantity for the formation of an ion pair from its cation and anion which are initially infinitely separated.

As an example, heat of dilution experiments and the information they provide about solution properties will be discussed. A comparison of Eqs. (12a, 22) and Fig. 1 shows that Φ_H^{rel} (in the literature sometimes called Φ_L) and the measured negative heat of dilution are identical, *i.e.*

$$\Phi_H^{rel} = -\Delta_{dil}H^\infty\,. \tag{24}$$

Hence, when diluting an electrolyte solution at molality m to a molality m′ by adding an appropriate amount of solvent, the accompanying heat of process is

$$\Delta\Phi_H^{rel} = \Phi_H^{rel}(m') - \Phi_H^{rel}(m) = \alpha'\Phi_H^{rel}(FI) - \alpha\Phi_H^{rel}(FI) + (\alpha - \alpha')\,\Delta H_A^0\,. \tag{25a}$$

Using the well-known thermodynamic relationship

$$\Phi_H^{rel}(FI) = -\nu RT^2\,\frac{1}{m}\int_0^m \left(\frac{\partial \ln \gamma'_\pm}{\partial T}\right)_p dm \tag{25b}$$

and the activity coefficient of the free ions required by the chemical model, $\gamma'_{\pm}$, after conversion of $y'_{\pm}$ (Eq. 7) into the molal scale, the following theoretical expression [77] is obtained

$$\Phi_H^{rel}(FI) = -\nu RT^2 \left\{ \left[\left(\frac{\partial \ln \varepsilon}{\partial T} \right)_p + \frac{1}{T} \right] \frac{\varkappa q}{1 + \varkappa R} + \frac{\bar{\alpha} \varkappa q}{3} \sigma(\varkappa R) \right\} \tag{25c}$$

where the function $\sigma(\varkappa R)$ is given by the relationship

$$\sigma(\varkappa R) = \frac{3}{(\varkappa R)^3} \left[1 + \varkappa R - \frac{1}{1 + \varkappa R} - 2 \ln (1 + \varkappa R) \right] . \tag{25d}$$

$\bar{\alpha}$ is the cubic expansion coefficient of the solvent; the other symbols where already defined in the preceding text.

The data analysis of dilution measurements with the set of Eqs. (25) yields the basic quantities K_A and R of the chemical model and ΔH_A^0 (heat of ion-pair formation). The entropy of ion-pair formation, ΔS_A^0, can be calculated from K_A and ΔH_A^0 in the usual way. Table III shows examples of a simultaneous determination of K_A, R_{exp} and ΔH_A^0.

The association constants of table III can be compared with those from conductance measurements, $K_A^{(\Lambda)}$, and are found to be in perfect agreement, e.g. $K_A^{(\Lambda)}(MgSO_4/H_2O) = 160\ dm^3\ mol^{-1}$. The agreement of the R_{exp}-values of Table III for aqueous solutions with those of the ion-pair model, Eq. (20), should be stressed as an important result. The calculated values, R_{calc}, correspond to $R = a + 2s$ (here $s = d_{OH}$, dimension of OH) according to this model. The agreement of R_{exp} with Bjerrum's distance parameter q, which is often used as the upper limit of association and which depends only on the permittivity of the solvent [cf. Eq. (19b)], is less satisfactory. For aqueous solutions of 2,2-electrolytes at 25 °C q equals 1.43 nm, independent of the ionic radii.

The same situation is given for non-aqueous solutions, e.g. propanol solutions in Table III. The association constants from calorimetric and conductance measurements

Table III Thermodynamic quantities of ion-pair formation in water and propanol (25 °C) from measurements of heats of dilution [77]

Solvent	Electrolyte	K_A / $dm^3\ mol^{-1}$	ΔH_A^0 / $J\ mol^{-1}$	R_{exp} / nm	$R_{calc/nm}$		
					$a + 2d_{OH}$	a + s	q
H_2O	$MgSO_4$	161	5780	0.93	0.88		1.43
	$CaSO_4$	192	6670	0.95	0.91		
	$CdSO_4$	239	8390	0.96	0.91		
	$NiSO_4$	210	5440	0.91	0.88		
C_3H_7OH	NaI	206,	18930	0.97	0.88	1.01	1.37
	KI	374	19060	0.99	0.91	1.04	
	RbI	527	17550	0.95	0.93	1.06	

agree satisfactorily [77]. The experimentally determined distances, R_{exp}, differ distinctly from q, which equals 1.37 nm for all propanol solutions of 1,1-electrolytes at 25 °C. Two values of R_{calc} are quoted in Table III, one for the configuration $C^+(OH)(OH)A^-$, *i.e.* $R_{calc} = a + 2d_{OH}$, the other for $C^+(propanol)A^-$, *i.e.* $R_{calc} = a + s$ (cf. Appendix A-2). Both values R_{calc} are compatible with the experimental value, R_{exp}.

Comparing Table III with the results of conductance data at various temperatures shows a further important feature. The temperature-dependence of conductance data yields ΔH_A^0-values *via* the relationship

$$\left(\frac{d \ln K_A}{dT}\right)_p = \frac{\Delta H_A^0}{RT^2} \tag{26}$$

which agree with the ΔH_A^0-values from calorimetric measurements, *i.e.* from conductance measurements on solutions of NaI in propanol a value of $\Delta H_A^0 = 18\,800 \text{ J mol}^{-1}$ is determined.

The second example concerning heat of solution measurements was chosen to stress a crucial problem in non-aqueous electrochemistry. This is the proper extrapolation to infinite dilution when association of the electrolyte occurs [84–87]. Figure 4 shows that the validity range of the limiting law is attained only at very low concentrations (here $<10^{-5}$ M), generally inaccessible to measurements. Hence, extrapolation from measured values ($>5 \cdot 10^{-3}$ M) yields erroneous data. Reliable

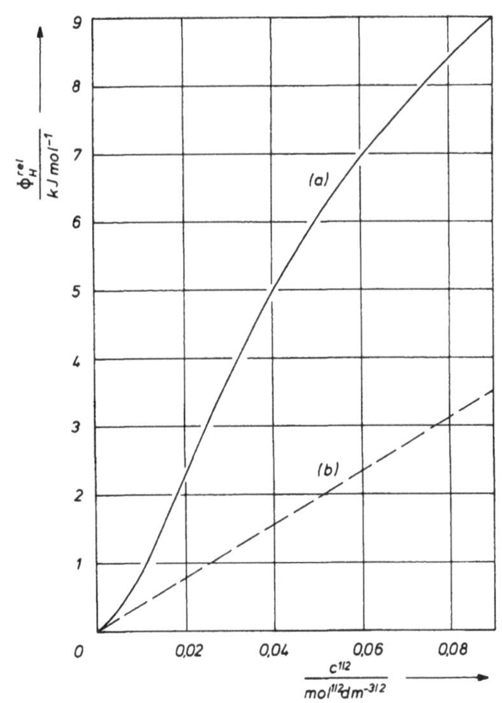

Fig. 4. Relative apparent molar enthalpy of KI in propanol (25 °C) from heat of dilution measurements. (**a**) measured curve; (**b**) limiting law; for explanation see text

Table IV Heat of solution data of NaI in 1-propanol at 25 °C [84]

$\frac{m \cdot 10^3}{mol\ kg^{-1}}$	$\frac{\Delta_{sol}H_Y}{J\ mol^{-1}}$	$\frac{\Phi_H^{rel}}{J\ mol^{-1}}$	$\frac{\Delta_{sol}H_Y^\infty}{J\ mol^{-1}}$
20.56	−19090	8898	−27988
21.70	−18850	9024	−27874
22.44	−18840	9103	−27943
22.64	−18780	9123	−27903
23.70	−18700	9250	−27950
		Mean:	−27932

extrapolations require the use of Eq. (23) and an appropriate model. The Debye-Hückel limiting law and its empirical extensions are generally insufficient.

Table IV shows $\Delta_{sol}H_Y$, Φ_H^{rel} and extrapolated $\Delta_{sol}H_Y^\infty$-values as a function of molality for solutions of NaI in n-propanol [84]. The Φ_H^{rel}-values were determined from the dilution measurements of Table III. Taking into account that Φ_H^{rel} contributes about 50% of $\Delta_{sol}H_Y$, it is obvious that reliable $\Delta_{sol}H_Y^\infty$-values can only be obtained from theoretically sound extrapolation methods. For comparison, Abraham et al. [85] estimated a value of $\Delta_{sol}H_Y^\infty = -23510\ J\ mol^{-1}$ from their measurements on solutions of NaI in propanol. A re-evaluation of these measurements based on the chemical model and Eq. (23) yields $\Delta_{sol}H_Y^\infty = -27200 \pm 180\ J\ mol^{-1}$, in satisfactory agreement with our values in Table IV.

An example of Eq. (23) for determining molar volumes is given in Ref. [81]. Comparing the results of molar volume measurements with the pressure dependence of the association constant from conductance experiments shows satisfactory agreement, *i.e.* the equation

$$\left(\frac{d\ln K_A}{dp}\right)_T = -\frac{\Delta V_A}{RT} \tag{27}$$

is fulfilled when K_A-values from pressure-dependent conductance and ΔV_A-values from density measurements are combined.

5.3 EMF-Measurements

For ΔG-measurements, emf or solubility products, the chemical potential, Eqs. (6, 7), is an appropriate form of the basic Eqs. (21). For example, the emf of the galvanic cell without liquid junction

$$Ag/AgCl(s)/K^+Cl^-\ (sol;\ concentration\ c\ mol\ dm^{-3})/(K, Hg)$$

is given by the set of equations

$$E = E^0 - \frac{2RT}{F}\ln(\alpha y'_\pm c) \tag{28a}$$

$$K_A = \frac{1-\alpha}{\alpha^2 c}\frac{1}{y'^2_\pm}; \qquad y'_\pm = \exp\left[-\frac{\varkappa q}{1+\varkappa R}\right]. \tag{28b, c}$$

Emf measurements yield reliable standard potentials E^0 only when data analysis uses well-founded extrapolation methods which take into account association of the electrolyte compounds [88]. A knowledge of reliable standard potentials is important for electrochemistry in non-aqueous solutions, especially for solvation studies and technological investigations. A comprehensive survey of these questions is in preparation. Data analysis with the help of Eqs. (28) gives K_A and R values which are compatible with those from other methods. Table V illustrates the satisfactory agreement of activity coefficients from emf measurements, $y_{\pm}$(emf), and heats of dilution, $y_{\pm}(\Phi_H^{rel})$, both evaluated by appropriate methods.

Table V Activity coefficients of aqueous $CdSO_4$ solutions at 25 °C

$\frac{10^4\,c}{\text{mol dm}^{-3}}$	10.0	30.0	50.0	100.0
$y_{\pm}(\Phi_H^{rel})$	0.699	0.551	0.476	0.383
$y_{\pm}$(emf)	0.698	0.552	0.481	0.388

5.4 Some Remarks on Thermodynamic Investigations

The possibility of determining the quantities K_A, ΔZ_A^0 and R controlling association as well as the values of μ_Y^∞ and $\Delta_{sol}Z_Y^\infty$ relevant for solvation gives a convincing reason for making comprehensive measurements on the properties of dilute solutions. Temperature- and pressure-dependent data in particular are needed at the present time. The determination of thermodynamic quantities has been the object of numerous investigations. However, the examples given illustrate the difficulties in getting standard values for the thermodynamic properties of partially associated electrolytes. This may be the reason why most of the work on the thermodynamic properties of electrolyte solutions, such as measurement of the apparent molal volumes [89–96], apparent molal heat capacities [89,90,97] and heats of solution [98–100], has been concerned with solutions of non-associated electrolytes. In this case the extrapolation to infinite dilution is carried out with the help of Eq. (12a) in combination with the Debye-Hückel activity coefficient or its extended forms [101–104].

The features of the chemical model are well-founded extrapolations towards Z_Y^∞ and related quantities on the one hand and the generation of a basic set of model parameters R and W_{ij} (*via* K_A) independent of the special experimental method on the other hand. Moreover, the distance parameter R is always found to be in accordance with the dimensions of a configuration of ions and orientated adjacent solvent molecules which is compatible with general chemical evidence. As far as chemical models are McMillan-Mayer-level Hamiltonian models, they permit the use of statistical thermodynamic relationships for calculating the solvent properties in a well-founded manner.

It should be mentioned in this context that investigations on dilute solutions, from which reliable information is expected, require precise measurements down to very low concentrations ($c \sim 10^{-4}$ mol dm^{-3}). Very sensitive and precise apparatus is needed, see Refs. [4,77,79,81,105–107], and purification of solvents and solutes together with the purity control of solvents, solutes and solutions are often the major part of these investigations.

5.5 Ion Solvation

Ion solvation is the transfer process of the separated ions of a pure electrolyte compound Y from the vacuum to the infinitely dilute solution in a solvent S. In the case of ionophoric electrolytes the solvation quantities $\Delta_{solv}Z_Y^{\infty}$ are related to the corresponding solution quantities, $\Delta_{sol}Z_Y^{\infty}$ (Eq. (12b)), via lattice quantities, $\Delta_{lat}Z_Y$ according to Fig. 1.

Lattice energies of many electrolytes are known [108–111] and in combination with the experimentally determined solution energies yield the solvation energies and related quantities $\Delta_{solv}Z_Y^{\infty}$. For tables of solution data see Refs. [36,112,113].

Besides the solvation quantities, transfer quantities, $\Delta_{tr}Z_Y^{\infty}$, can be advantageously used, cf. Fig. 1. They give an account of the change in Z when the electrolyte Y is transferred from solvent S′ to solvent S. With water as the reference solvent, S′, the transfer activity coefficients, ${}_m\gamma_Y$ [114–116], are obtained from the Gibbs transfer energy, $\Delta_{tr}G_Y^{\infty}$, by the relationship

$$\Delta_{tr}G_Y^{\infty} = \Delta_{solv}G_Y^{\infty}(S) - \Delta_{solv}G_Y^{\infty}(W) = RT \ln {}_m\gamma_Y \,. \tag{29}$$

The choice of the appropriate concentration scale for standard thermodynamic functions of transfer was extensively discussed by Ben-Naim [117] who showed that the molarity scale has a number of advantages over the others.

Separation into ionic transfer activity coefficients for the electrolyte $C^{z+}A^{z-}$ is executed with the help of the equation

$${}_m\gamma_Y^2 = ({}_m\gamma_+)\,({}_m\gamma_-) \,. \tag{30}$$

The interest of theory and technology in single ion solvation and transfer quantities originates in their importance for solution structure, kinetic, analytical or surface problems, *i.e.* for all problems involving the solvation shell or its rearrangement. For example, transfer proton activity coefficients ${}_m\gamma_{H^+}$ are used for transferring the pH-scale from water to other solvents:

$$pH\,(S) = pH\,(W) + \log {}_m\gamma_{H^+} \,. \tag{31}$$

Further examples of technical importance are found in the field of extraction processes [118], ionic equilibria and emf measurements [119], and analytical applications [120,121].

The requirements of theory both for solvation and transfer data of single ions are similar. A complete theory would require the knowledge of all molecular distribution functions and mean-force potentials between the ions and the solvent molecules. As already stressed in Section II such a theory is unavailable with the present state of knowledge. In the endeavour to represent solvation by models, the ion-solvent interactions are split into electrostatic, non-electrostatic, and chemical contributions.

Dielectric continuum models like the Born model [122] and its extensions taking into account local permittivity [123, 124] or introducing 'effective ionic radii' [125, 126] consider the solvent as a structureless shielding continuum. More satisfactory results are obtained with the help of an electrodynamic theory of condensed media which takes into consideration the structure of polar solvents [127], but the prediction of equal free energies of hydration for cations and anions of equal size and charge is rather unrealistic. These quantities should differ by twice the ion-quadrupole interaction energy.

Ion-solvent interaction causes orientation of the neighbouring inner solvent molecules and extends with greater or less attenuation into the bulk solution [128–131]. Primary, and in some cases also secondary, solvation shells are chosen as the basis of models. Solvent mixtures introduce the possibility of preferential ion solvation [132–137].

The energy content of the primary solvation shell depends on the short-range interactions of the types of ion-dipole, ion-quadrupole, ion-induced dipole, dipole-dipole and dispersion and repulsion forces [138–140]. For non-aqueous electrolyte solutions, however, most of the molecular and structural data needed for the calculation of the formation energy of the primary solvation shell are unknown. In addition, calculations of solvation energies must include an estimation of the energy content of the region outside the primary solvation shell. Whereas the electrostatic part may be approximated by the Born model, using the solvated ion as a hard sphere, the non-electrostatic parts resulting from cavity formation and structure breaking in the secondary solvation shell are generally unknown. Therefore, ab initio calculations have not so far been very successful for non-aqueous solvation. Present information has been obtained from semi-empirical methods and/or extra-thermodynamic assumptions.

The basis of the extrapolation methods [141–143] is the representation of the electrostatic contributions to ion-solvent interactions by a series expansion of type $a_j r^{-j}$ where r is the ionic radius. For example, in a series of electrolytes with common cation and varying anions (anion radius, r_-) a set-up

$$\Delta_{sol}Z_Y^\infty = \Delta_{sol}Z_+^\infty + \sum_{j=1}^{n} \frac{a_j}{r_-^j} + \Delta_{sol}Z_-^\infty(\text{neutral}) \tag{32}$$

is used for determining single cation solvation quantities from the measured $\Delta_{sol}Z_Y^\infty$ values. A variety of methods differ in the number of terms in the power series and in the corrections, $\Delta_{sol}Z_-^\infty$(neutral), for non-electrostatic contributions [144–146]. A drawback of all extrapolation methods is the uncertainty of the extrapolation ($r \to \infty$) from the very small range of ionic radii available.

Reference ions [147], reference ion/molecule pairs [148–150], and reference electrolytes [91, 136, 150–156], are the basis of further methods. The assumption that the rubidium ion would have a constant potential in all solvents as a consequence of its low electrical field and polarizability [147], *i.e.* ${}_m\gamma_{Rb^+} = 1$, as well as similar hypotheses involving large transition state anions [149] are only of historical interest. More realistic are the determinations of solvation quantities based on the assumption that large ions and their corresponding uncharged molecules, like the ion/molecule pair ferrocinium/ferrocene [148] or similar ones [149,150], show equal solvation in each solvent. The disadvantage of the reference ion and reference ion/molecule methods, both of which neglect the electrostatic contributions of the ionic species, is assumed to be overcome [151] by the reference electrolyte method proposing the partition of measured electrolyte solvation quantities equally between its ions. The criteria for reference electrolytes are the same as for reference ion/molecule pairs. The difference in the short-range electrostatic ion-molecule interactions is neglected if the cation and the anion of the reference electrolyte are structural analogues of equal size, e.g. $(Ph_4As)^+(BPh_4)^-$ [151] or $(i\text{-}Am_3BuN)^+(BPh_4)^-$ [150]. Reference electrolytes were widely used to determine transfer activity coefficients and solvation enthalpies [111,112,131,150,153] and partial molar volumes [91,154] of single ions. Criticisms of the reference electrolyte method are based on spectroscopic investigations [157] and calculations [158] using the scaled-particle theory [159]. Kim et al. [155,156] propose an unequal partition of the standard free energy of transfer between $(Ph_4As)^+$ and $(BPh_4)^-$ as the result of a comprehensive investigation. However, the consequences of these corrections for the equipartition principle are still a matter of debate. In a recent review Marcus [119] gives a critical evaluation of the $(Ph_4As)^+(BPh_4)^-$ problem.

A rapid experimental method to determine transfer activity coefficients uses galvanic cells with transference but negligible liquid-junction potentials [149,160–162], e.g. the cell

Ag	$AgClO_4$ (0.01 M)	Et_4NPi (0.1 M)	$AgClO_4$ (0.01 M)	Ag
	reference	*reference solvent*	*solvent S*	
	solvent	*or solvent S*		

where the reference solvent is either acetonitrile or methanol and the bridge electrolyte tetraethylammonium picrate dissolved in the weaker solvating solvent for the Ag^+ ions. The results from the application of this method differ by about one unit in $\log {}_m\gamma_Y$ from those obtained by the reference electrolyte assumption which Popovych [114] considers to be the more reliable one.

All these methods do not satisfactorily take into account such specific interactions like H-bonding of small anions with protic solvents [163] or those of Cu^+, Ag^+, Au^+ with some dipolar aprotic solvents, as well as the structural contributions of strongly structured solvents, e.g. water, alcohols, amides, to the solvation energy. This drawback of actual solvation theory is one of the reasons motivating practical chemists to restrict themselves to qualitative classification [164,165] according to solvent scales or classes, cf. also Section 8.3.

5.6 Concentrated Solutions

Information in the field of thermodynamic properties of concentrated non-aqueous electrolyte solutions is rather poor at present, in spite of increasing technological interest. No reliable theory is available. Aqueous solutions have been better investigated. Pitzer et al. [166–171], and Cruz and Renon [172–174] extended the Debye-Hückel concept and obtained empirical equations for the osmotic coefficient and ionic activity coefficients from Gibbs excess enthalpy which in Pitzer's theory is presented as a sum of cationic and anionic contributions with an unlimited number [174] of adjustable parameters. The Cruz-Renon extension contains a Debye-McAulay term to correct for the change of dielectric constant of the solvent due to the presence of ions in the solution and a term taking into account short-range forces. Both theories are of great help for practising engineers in the prediction of solubility data, treatment of vapour-liquid equilibria, and further technical problems [174].

5.7 Water at Extreme External Conditions

Application of high pressures and temperatures changes the bulk properties of a solvent to a considerable extent. There is a continuous transition of these properties going through the critical point [175]. As emphasized by Franck [176], under extreme conditions water substance loses its typical properties and behaves as a high polar "non-aqueous, water-like" solvent. The consequences for electrolyte solutions are considerable changes in the ion-ion and ion-solvent interactions. Thus it is not surprising that 1,1-electrolytes form ion pairs in aqueous solutions at high temperatures, e.g. aqueous NaCl solutions with densities below 0.75 g cm^{-3} exhibit ion-pairing, increasing with decreasing density [177]. For further information on this technically important solvent and its "non-aqueous" electrolyte solutions, including conductance and dielectric properties see Refs. [178-188].

VI Transport Properties

6.1 Dilute Solutions

Transport equations of electrolyte and single ion conductance, self- and mutual-diffusion, and transference numbers can be obtained either from Onsager's continuity equation or from Onsager's fundamental equations of irreversible processes. Many publications deal with this matter, especially with electrolyte conductance. For monographs, review articles, surveys of results and recent contributions in this field see Refs. [30,36,44,46,47,52,53,67,69,76,82,175,189-200]. Recent extensions of conductance theory concern electrolyte mixtures of any type [201-204]. A discussion of former papers is given in Refs. [36,76,189-191].

Electrolyte conductance, Λ, and transference numbers, t_i, are required for a proper understanding of the transport of charge by ions in electrolyte solutions.

Conductance equations for completely dissociated electrolytes are obtained in the form

$$\Lambda = \Lambda^{\infty} - \Lambda^{r} - \Lambda^{e} \tag{33}$$

with Λ^{∞} as the limiting electrolyte conductance at infinite dilution, a concentration-dependent relaxation term, Λ^{r}, and a concentration-dependent electrophoretic term,

Λ^e, both of which are due to effects inhibiting the conductance as the electrolyte concentration increases. Alternatively, series expansions

$$\Lambda = \Lambda^\infty - Sc^{1/2} + Ec \log c + J_1 c + J_2 c^{3/2} + \ldots \tag{34}$$

in which the coefficients S, E, J_1, and J_2 contain contributions due both to the relaxation and electrophoretic effects can be used. Summarizing tables of the coefficients as obtained from various theories are given in Refs. [36,191]. Eq. (34) contains the well-known square-root law

$$\Lambda = \Lambda^\infty - S\sqrt{c} \tag{35}$$

as the limiting law. The limiting law has been the object of various empirical extensions (see [44,47,191]) since Onsager derived it as the first exact and statistically well-founded conductance equation [45,205].

Only the free ions in the solution are supposed to transport charges in the applied external field. For associating electrolytes Eq. (34) is transformed into the set of equations

$$\Lambda = \alpha[\Lambda^\infty - S(\alpha c)^{1/2} + E(\alpha c)\log(\alpha c) + J_1(\alpha c) + J_2(\alpha c)^{3/2} + \ldots] \tag{36a}$$

$$K_A = \frac{1-\alpha}{\alpha^2 c} \times \frac{1}{y_\pm'^2}; \qquad y_\pm' = \exp\left[-\frac{\varkappa q}{1+\varkappa R}\right] \tag{36b, c}$$

which is of the type of Eqs. (21). Then the introduction of the chemical model as the base for electrolyte conductance is straightforward [4,74–76]. Discussion and comparison of conductance equations of electrolytes and single ions and their use for determining transport and structure parameters of a multitude of solutions can be found in recent surveys [4,67,75,76,189–192,194]. Fig. 6 shows the generally observed dependence on concentration of equivalent conductance of partially associated electrolytes. The validity range ($<10^{-5}$ M) of Onsager's limiting law, Eq. (35), is indicated. The deviations from the limiting law are caused both by the complete transport equation of type (36) (here from Ref. [195]) and by association.

Conductance measurements on dilute solutions are of special interest for electrolyte theory. These measurements can be carried out at high precision for almost all electrolytes in almost all solvents at various temperatures and pressures and thus provide an efficient method for determining the basic data of electrolyte solutions, *i.e.* Λ^∞, K_A and R, under various conditions. Values of K_A and R are found to be compatible with the values obtained from thermodynamic methods. The enthalpies and volumes of ion-pair formation, ΔH_A^0 and ΔV_A, as determined from temperature- and pressure-dependence of conductance, are compatible with the corresponding relative apparent molar quantities, $\Phi_H^{rel}(IP)$ and $\Phi_V^{rel}(IP)$, from thermodynamic measurements, cf. Section 5.2.; R-values are found to be almost independent of temperature.

Transference numbers, t_i, and single ion conductances, $\Lambda_i = t_i\Lambda$, deduced from them are of theoretical and technical interest. Several attempts have been made to replace the experimentally difficult and time-consuming measurement of transference numbers by methods based on empirical assumptions, such as splitting the equivalent conductance Λ of standard electrolytes into equal contributions from cation and anion (cf. Sect. 5.5.), e.g. $(Bu_4N)^+(BPh_4)^-$ [206,207] or $(i\text{-}Am_4N)^+(i\text{-}Am_4B)^-$ [208], or calculating the ionic conductance in one solvent from that in another by the help of Walden's rule [209]. The temperature-dependence of transference numbers was also calculated on the basis of rules of this type [210].

Electrolyte theory relates transport numbers to the chemical model through the following set of equations [82,194]

$$\frac{t_i - 0.5}{t_i^\infty - 0.5} = \Lambda^\infty \left[\Lambda^\infty - \frac{1}{\Xi} \frac{e_0^2 N_A}{3\pi\eta} \frac{\varkappa}{1 + \varkappa R} \right]^{-1} + B(\alpha c) \tag{37a}$$

$$K_A = \frac{1-\alpha}{\alpha^2 c} \times \frac{1}{y_\pm'^2}; \qquad y_\pm' = \exp\left[- \frac{\varkappa q}{1 + \varkappa R} \right] \tag{37b, c}$$

where e_0 is the charge of proton, N_A Avogadro's number, η the viscosity of the solvent and Ξ a conversion factor ($\Xi = 1$ for SI units). The other symbols have their usual meaning.

Among the experimental methods for determining transference numbers, the moving boundary method (cf. [82,190,194,211]) allows their determination with a precision close to that of conductances. Figure 5 gives an example [82].

Figure 5 shows transference numbers decreasing with increasing concentration and decreasing temperature when $t_+^\infty < 0.5$ and increasing when $t_+^\infty > 0.5$, in accordance with theory. The possibility of reversing the sign of temperature- and concentration-

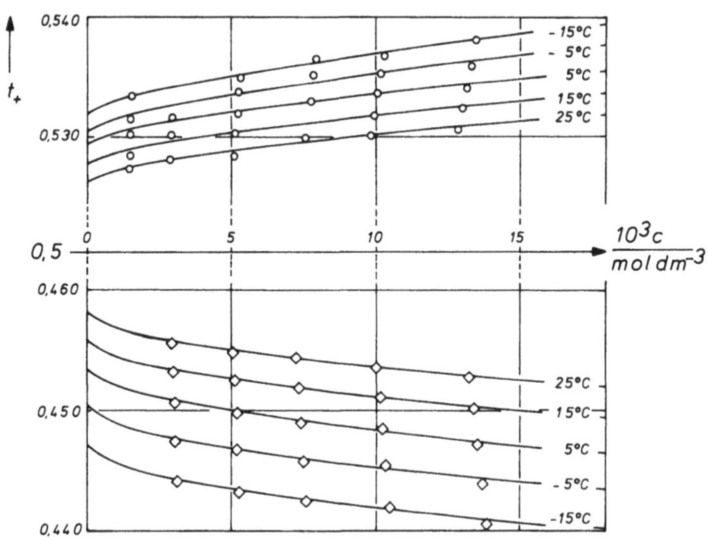

Fig. 5. Temperature- and concentration-dependence of cationic transference numbers of methanol solutions of Me_4NSCN (○) and KSCN (◇) [82]. The full lines are computer plots according to the set of Eqs. (37)

dependence by the choice of appropriate electrolytes can be used in battery research when battery electrolytes with a high mobility of the active ion at low temperature are sought as well as in other technological applications which require specific charge transport properties, cf. Part B.

Table VI shows the limiting anion conductances determined from the measurements underlying Fig. 5. Extrapolations were made by taking into account the association of the electrolyte compounds KSCN and Me_4NSCN in methanol *via* Eqs. (37). As a result, the theoretical requirement of equal limiting anion conductances $\lambda^{\infty}_{SCN^-}$ is fulfilled within the limits of experimental error.

For further information on transport phenomena in dilute solutions reference should be made to the recent literature; the literature up to 1975 is quoted in Refs. [36, 55, 113].

The current state of knowledge in the field of dilute electrolyte solutions is illustrated by Fig. 6 [72].

The conductance functions for Pr_4NI and i-Am_4NI in propanol, Fig. 6, are computer plots which were calculated entirely without use of conductance data for

Table VI Single anion conductance from independent measurements of transference numbers in KSCN and Me_4NSCN in methanol [82]

Temp. °C	$\lambda^{\infty}_{SCN^-}$/[S cm^2 mol^{-1}] from KSCN	$\lambda^{\infty}_{SCN^-}$/[S cm^2 mol^{-1}] from Me_4NSCN
−15	33.65	33.61
− 5	39.84	39.82
+ 5	46.57	46.60
+15	53.93	53.97
+25	61.87	61.95

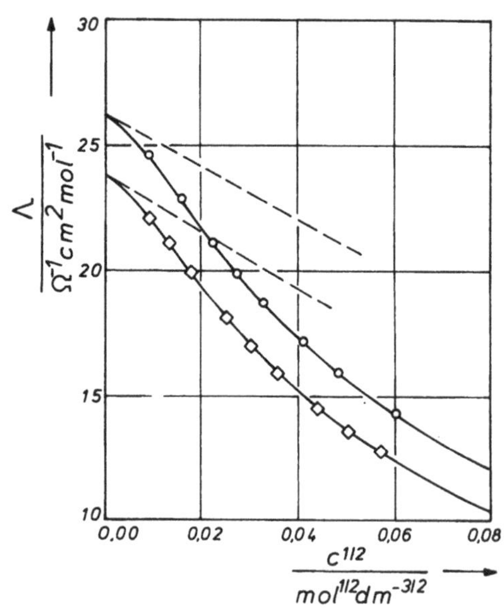

Fig. 6. Conductance of propanol solutions of Pr_4NI (○) and i-Am_4NI (◇) at 25 °C [72]. The full lines are calculated with the help of Eqs. (36) using λ_i^{∞}-values from transference numbers, and K_A and R values from calorimetric measurements. Values (○) and (◇) from conductance measurements were added for comparison. The broken lines represent the Onsager limiting law of conductance

the electrolyte solutions investigated. Calculation was based on the set of Eqs. (36). Association constants, K_A, and distance parameters, R, were determined from calorimetric measurements on solutions of Pr_4NI and i-Am_4NI in propanol[77]; Λ^∞-values were obtained via transport numbers t_i^∞ of KSCN[82] and single-ion limiting conductances λ_i^∞ based on these. The measured points of independent conductance measurements are added to prove the validity of the method. Agreement is better than 0.1%.

6.2 Concentrated Solutions

Theoretically well-founded equations for transport data of highly concentrated electrolyte solutions are not available at present. Three classes of transport equations can be found in the literature[76]; molten salt approaches, empirical extensions of equations for dilute solutions, and empirical equations for fitting measured data.

The temperature-dependence of transport properties W(T), of glassforming liquids and of fused salts[212] can be interpreted with the help of a modified Vogel-Fulcher-Tammann equation

$$W(T) = AT^{-1/2} \exp\left[-\frac{B}{T - T_0}\right] \qquad (38)$$

where A and B are constants for a given transport property, conductance, diffusion or fluidity. The theoretical significance of the ideal[213] or theoretical[214] glass transition temperature T_0 is given in Refs.[215] and[216]. Application of Eq. (38) to experimental results usually yields temperatures T_0 which are lower than the glass transition points T_g determined by thermal methods. Thermal methods, e.g. DSC or DTA, yield values of T_0 which depend on the cooling rate[217,218] and which cannot be attributed to internally equilibrated liquids. Angell[213] transformed Eq. (38) into a relationship for the isothermal concentration-dependence of transport properties predicting a maximum of specific conductance at each temperature, e.g. for aqueous $Ca(NO_3)_2$ solutions[213,218]. Bruno and Della Monica[219,220] have applied Angell's model to non-aqueous solutions. Further information can be found in comprehensive reviews[175,221-224].

Extensions of the conductance equations for dilute solutions using empirical correction factors, e.g. viscosity functions[225,226], can be found in the literature. However, although viscosity is the most important parameter influencing the conductance[227], equations of this type are doubtful. The mean-force potentials of dilute solutions are based on approximations which cannot really be corrected for by viscosity functions. A noteworthy extension is given in Ref.[228].

Comprehensive experimental data which provide an insight into the competitive effects of solvent viscosity, ion-ion, and ion-solvent interactions are only available at present for a few non-aqueous systems. A survey is given in Refs.[76,229,230].

Fig. 7 shows the typical concentration-dependence of the specific conductance, $\varkappa[\Omega^{-1}\,cm^{-1}]$, of $LiClO_4$ at various temperatures in propylene carbonate[232] and propylene carbonate-dimethoxyethane mixtures[227].

Lacking a reliable theory, an empirical equation[231] was used to reproduce the

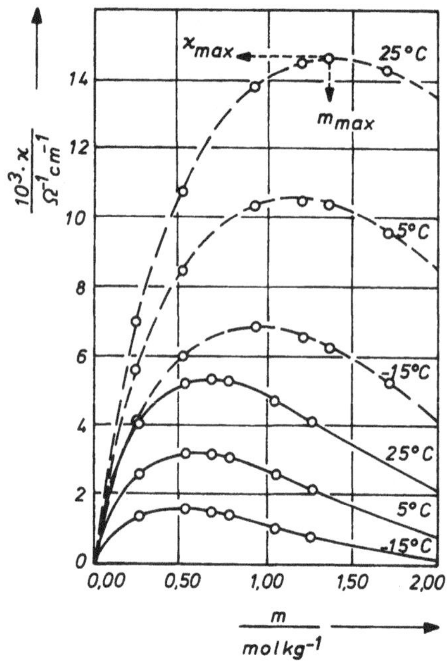

Fig. 7. Specific conductance of $LiClO_4$ solutions in propylene carbonate (full lines) and propylene carbonate — dimethoxyethane (42 weight % of PC) mixtures (broken lines) at various temperatures

conductance data. The concentration, m_{max}, at maximum conductance, $\varkappa_{max}$, decreases with decreasing temperature, *i.e.* with increasing viscosity of the solvent. A decrease of viscosity at constant temperature resulting from a change in solvent composition produces an increase in m_{max}, also proving that viscosity is the most important factor determining the conductance. It is of technical interest that the specific conductance of solutions in polar solvents, here PC, can be considerably increased by the addition of suitable solvents with low permittivity, here DME. Ion-ion and specific ion-solvent interactions, e.g. association and solvation, are second-order effects [227)].

Investigations on various salts in these solvents show that competition between ion-ion and ion-solvent interactions with increasing salt concentration leads to comparable states of the different solutions at their respective maximum conductance [227, 232)], *i.e.* the maximum specific conductance for every solution is obtained when the conductance-determining effects have established a critical energy barrier which depends almost exclusively on solvent and temperature and not on the solute. As an important consequence, in spite of large ion-ion association constants, the maximum specific conductance can be found at unexpectedly high $\varkappa_{max}$-values in solvents of low permittivity [227, 230)]. This feature can be used advantageously for technical applications.

6.3 Ion Aggregates and their Role in Conductance

Many 1,1-electrolytes show association to aggregates higher than ion pairs in pure solvents of classes 6 and 7 of Table I and in their mixtures of low permittivity, also in mixtures with solvents of other classes, e.g. PC, H_2O, ROH. The dependence of the

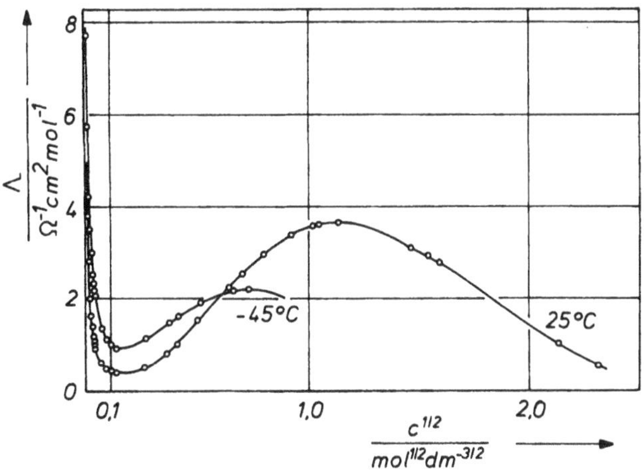

Fig. 8. Equivalent conductance of $LiBF_4$ in dimethoxyethane solutions at +25 °C and —45 °C [233]

electrolyte conductance of these solutions upon the concentration and temperature is a suitable probe for this effect. Fig. 8 shows a minimum of equivalent conductance followed by a maximum for DME solutions of $LiBF_4$ at 25 °C as a typical proof for the coexistence of free ions, ion pairs, triple ions and possibly higher aggregates in the solution [233]. The use of appropriate conductance equations [30, 234] for data analysis yields the equilibrium constants K_A (ion pairs) and K_T (triple ions).

The ion-pair constant K_A increases strongly with increasing temperature whereas the triple ion constant K_T decreases slightly. Taking into consideration the resulting change in the charge density and the weakly temperature-dependent viscosity of the solvent (cf. Fig. 18), the observed inversions of the temperature coefficient of equivalent conductance in the range of low and moderate concentrations can be predicted without the need for further assumptions [233].

The range of high concentrations of such solutions, especially of those with solvents of class 7, shows examples of an increase of conductance up to five orders of magnitude which cannot be explained by the usual transport equations. On the basis of conductance [235, 236], viscosity [235], transference numbers [237] and cryoscopic [238] measurements Gileadi et al. considered a "hopping mechanism" [235, 236] quite similar to the Grotthuss mechanism for the proton in aqueous solutions

$$C^+ + (C^+A^-)_n \rightarrow C^+ \dots A^-(C^+A^-)_{n-1} \dots C^+ \rightarrow (C^+A^-)_n + C^+$$

which predicts an increase of conductance with increasing number n, *i.e.* with increasing ion aggregation in the solution. The role of both cluster formation and the "hopping mechanism" for solutions of low dielectric constant is also illustrated by the system NMF/Br_2 [239] in which dissociation occurs according to

$$NMF + Br_2 \rightarrow (NMFBr^+)Br^- \xrightarrow{Br_2} NMFBr^+ + Br_3^- .$$

Ion aggregates of quite another type must be considered for polyvalent ions in non-aqueous solvents, even in protic solvents (class 1), e.g. $CdCl_2$ in methanol [240]. The concentration-dependence of equivalent conductance down to 10^{-5} M solutions as well as the temperature-dependence require this electrolyte to be treated as a 1,1-electrolyte, $CdCl^+Cl^-$, forming ion pairs of type $[CdCl^+(ROH)Cl^-]$.

VII Spectroscopic and Relaxation Methods

7.1 Introduction

The information on the structure of electrolyte solutions provided by thermodynamic and transport properties on the one hand and by spectroscopic, relaxation and kinetic investigations on the other, complement one another with regard to the chemical model. Thermodynamic and transport properties provide the distance parameter R, the overall association constant K_A, and the activity coefficient $y_\pm$ linked to it. No direct information can be achieved on the structure of the region $a \leqq r \leqq R$ and possible regions $a \leqq R_1 \leqq R_2 \dots \leqq R$. This problem, however, can be solved by modern spectroscopic and relaxation methods.

An appropriate subdivision of the region $a \leqq r \leqq R$ (cf. Sect. 4.2.) within the framework of an equilibrium concept is accompanied by the assumption of a suite of equilibria [74, 75]

$$C^{z+} + A^{z-} \overset{K_1}{\rightleftarrows} [C^{z+}A^{z-}]_1^0 \overset{K_2}{\rightleftarrows} \dots \overset{K_n}{\rightleftarrows} [C^{z+}A^{z-}]_n^0 \tag{39}$$

Equation (20) gives an example. Every "ion pair" $[C^{z+}A^{z-}]_i^0$ differs as a chemical species by its energy content from its neighbours in the suite as a consequence of the distance-dependent interaction forces between the ions. Individual equilibrium constants are given by the relationships

$$K_1 = \frac{a_1}{a_+ a_-}; \qquad K_i = \frac{a_i}{a_{i-1}} \qquad (i = 1 \dots n) \tag{40}$$

a_+ and a_- being the activities of the free ions and a_i those of the ion pairs.

Actually the link between spectroscopic, on the one hand, and thermodynamic or transport process results on the other, is not yet completely established. The concentration ranges of the methods differ greatly. Spectroscopy requires moderately or highly concentrated solutions with the exception of a few methods which have recently been developed for the low concentration range. Hence, the use of activity coefficients $y'_\pm$ of dilute solutions is doubtful. The overall association constant

$$K_A = K_1 + K_1K_2 + K_1K_2K_3 + \dots + \prod_{i=1}^{n} K_i \tag{41}$$

considers exclusively free ions and ion pairs $[C^+A^-]_i^0$. Only in dilute solutions the activities of the ion pairs can be set equal to their concentrations.

The overall constant of the chemical model (see Fig. 3)

$$K_A = 4000\pi N_A \sum_{i=1}^{n} \int_{R_{i-1}}^{R_i} r^2 \exp\left[\frac{2q}{r} - \frac{W_{+-}^{*(i)}}{kT}\right] dr \tag{42}$$

when compared with Eq. (41) enables the partial association constants to be expressed in terms of R_i and $W^{(i)}_{+-}$ [3]. Hence, distance parameters R_i and ion-pair configurations $[C^+A^-]^0_i$ from spectroscopic evidence as well as information on ion-ion and ion-solvent interaction in moderately and highly concentrated solutions are actually the crucial contributions to electrochemistry.

Relaxation methods can provide information on some of the equilibria of Eq. (39) or all of them, depending on the frequency range of the method used, and can then establish a criterion for an appropriate subdivision of the region $a \leqq r \leqq R$.

Diffraction methods yield direct information on the pair-correlation functions, $g_{ij}(\vec{r}_1, \vec{r}_{21})$, cf. Section 4.1.

Absorption methods give evidence on ion pairs but do not necessarily yield the overall association constants required by the thermodynamic excess functions. However, they furnish essential contributions to the elucidation of solvent-solvent and ion-solvent interactions and are the basis for most of the semi-empirical interaction scales usually applied.

7.2 Diffraction Methods

X-ray and neutron scattering experiments yield direct information on the atomic pair-correlation functions, $g_{ij}(\vec{r}_1, \vec{r}_{12})$, by the help of the scattering cross-section $d\sigma/d\Omega$ [241, 242] or the reduced intensity I [243] both of which depend on the scattering angle θ, which is expressed as the scattering vector $k = [4\pi \sin(\theta/2)]/\lambda$. λ is the wavelength of the incident radiation. The intensity of the radiation scattered by each atom i depends on its scattering factor f_i.

X-rays are scattered at the inner electrons of an atom. The scattering factor (also called form factor) decreases with increasing k and increases linearly with the atomic number. The positions and interactions of light particles like hydrogen atoms or protons cannot be determined directly. Thermal neutrons, in contrast, are scattered on the spherical atomic nuclei and therefore the scattering factor (in this case also called coherent scattering length) is independent of angle but shows marked, unsystematic dependence on the atomic mass for all nuclei. The elimination of systematic errors due to absorption, polarisation, sample geometry, Compton scattering, multiple scattering at many atoms, incoherent and inelastic scattering, is a difficult but necessary task.

The mean cross-section is a function of the total structure factor $F_T(k)$

$$\frac{d\sigma}{d\Omega} = N \times [F_T(k) + \Sigma\, c_i f_i^2]\,, \tag{43a}$$

which itself is expressed by the partial structure factors $S_{ij}(k)$

$$F_T(k) = \sum_i \sum_j N c_i c_j f_i f_j\, [S_{ij}(k) - 1] \tag{43b}$$

with

$$S_{ij}(k) = a + \frac{4N}{kV} \int_0^\infty (g_{ij}(r) - 1) \times r \times \sin(kr)\, dr\,; \qquad \sum c_i = 1 \tag{43c, d}$$

N is the total number of all particles i with concentration c_i in the volume V.

The total structure factor $F_T(k)$ contains the simultaneous information on all correlation functions of adjacent atoms, intermolecular as well as intramolecular. As a consequence, reports in the literature have been limited to solvents with few atoms, e.g. H_2O. Furthermore, the measured signal is very insensitive towards changes in solvent structure so that only highly concentrated solutions can be examined. The important results obtained from diffraction methods are the $M^{z+} \dots OH_2$ distances, the tilt angles between the $M^{z+} \dots O$ axis and the plane of the water molecule and hydration numbers [241, 242]. The configuration of $CdI_{4-n}^{(n-2)} \cdot S_n$ complexes (S = H_2O or dimethylsulfoxide) were investigated recently [243].

The extended X-ray absorption fine structure (EXAFS) in the high energy vicinity of the K-edge (in some cases also L-edge) of atoms [244, 245], e.g. Cu, Ni, Fe, Br, results from the backscattering of the excited photo-electron at the electron shells of the nearest neighboured atoms. The distance-dependent interference between the outgoing and the back-scattered wave function of the photo-electron changes the interaction between the incident X-ray radiation and the excited photo-electron and causes the fine structure of the transmitted radiation. The selectivity of the absorption edge, the dependence of the shape of the fine structure on the atomic number of the backscattering atom, the absence of multiple scattering, and the ability to measure the phase and amplitude of the back-scattered radiation allow a rather simple and straightforward interpretation of EXAFS [244]. Data analysis *via* Fourier transformation provides very exact ion-solvent and ion-ion distances in moderately concentrated as well as in dilute solutions. The main advantage, but also the main experimental difficulty, is the need for continuous synchrotron radiation with high energy flux.

Until now most of the investigations have been made on aqueous solutions, e.g. $CuBr_2$ [246, 247], $NiCl_2$ [248], $Ni(NO_3)_2$ [249]. The chemical behaviour of these metal

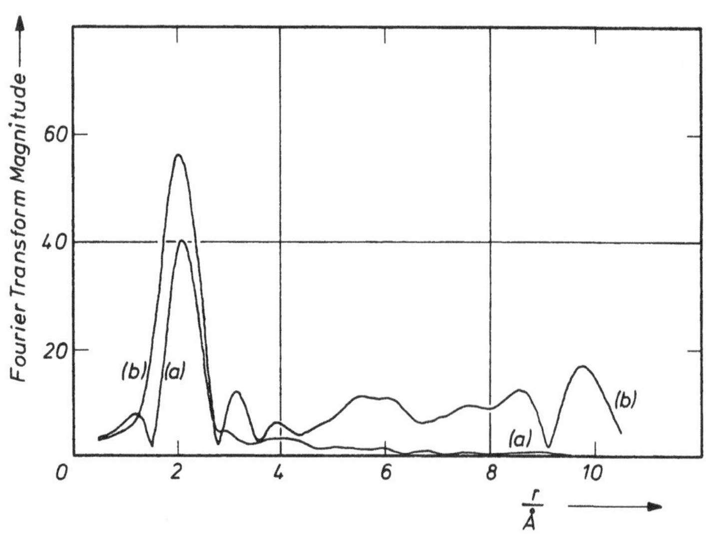

Fig. 9. Fourier transform EXAFS of $Ni(NO_3)_2 \cdot 6\,H_2O$ (**a**) solid salt; (**b**) 0.1 M aqueous solution (with kind permission of the authors [249])

cations in forming stable metal-solvent complexes is clearly shown by the investigations on $Ni(NO_3)_2$ for which the distances and the coordination number are the same in 0.1 M aqueous solutions and in $Ni(NO_3)_2 \cdot 6\,H_2O$ crystals [249] as shown in Fig. 9.

Investigations on moderately concentrated solutions (0.1 M) of $NiCl_2$ [248] reveal outer-sphere complexes; the existence of $CuBr_4^{2-}$ [246] or $NiCl_4^{2-}$ [248] could not be verified by this method. An example in non-aqueous solution, LiBr in diethyl ether, has recently been published [250,251].

The consistency and the accuracy of the data for intermolecular distances and coordination numbers are the main features of these three methods. Their wide application, however, is prevented by the very expensive and time-consuming nature of the experiments.

7.3 Absorption Spectroscopy

In contrast to diffraction methods, yielding the position variables and quantities related to them, absorption spectroscopy also furnishes information about the population density of energy levels via Beer's law.

Electronic spectroscopy (180–750 nm) of organic compounds with mesomeric systems, e.g. $CH_3—CO—CH=C(CH_3)_2$ [252], or appropriate inorganic metal complexes, e.g. $CoCl_2(C_5H_5N)_n$ [253], shows marked solvent effects from $\pi \rightarrow \pi^*$, $n \rightarrow \pi^*$, $d \rightarrow \pi^*$ transitions or charge-transfer absorptions. Organic ions without mesomeric systems, simple inorganic ions and most of the solvents in Table I are generally unsuited for investigations in this frequency range. Halogen ions are an exception to this rule. They show $p \rightarrow s$ and $p \rightarrow d$ transitions in the UV-range which are shifted to lower energies for the higher homologues thus making iodide the most convenient ion for such measurements [254]. The absorption bands, which are strongly solvent-dependent, represent charge transfer to solvent (CTTS) spectra.

Investigations on carbanions with delocalized electrons are an interesting field of application [255,256]. Their spectra in solvents of low permittivity suggest equilibria between tight and solvent-separated ion pairs and higher aggregates. For example, ion pairs formed from the alkali salts of isomeric vinyl pyridine carbanions [257] show the marked influence of the solvent and the structure of the anion as a consequence of charge delocalization in the isomeric pyridine rings. The properties of such carbanion-cation pairs, which serve as catalysts in non-radical polymerization, determine to a large extent the stereospecificy of the chain growth and the stereoregularity of the polymer [258].

UV and VIS spectroscopy provide a handy and efficient tool for the generation of semi-empirical scales of solute-solute and solute-solvent interaction parameters which are currently used for correlations. Table VII gives a survey.

Excited states of ketones (compound I) and positive solvatochromic dyes, e.g. diethylamino p-nitrobenzene (compound II of table VII) or related compounds are more strongly solvated by polar solvents when compared with apolar ones. The shift of the $\pi \rightarrow \pi^*$ transition of substituted p-nitrophenoles [260] as well as the $n \rightarrow \pi^*$ transition of ketones [259] are used as empirical polarity scales, π^* [260] and Φ [259], of a solvent. The energy shifts of the $\pi \rightarrow \pi^*$ transition of substituted pyridinium-N-phenolate betaines (compound III) and the charge-transfer absorption

Table VII Selected compounds for UV/VIS-solvent parameters

No	Compound	Solvent parameter	References
I	$CH_3\overset{O}{C}CH_3$	Φ	[259]
II	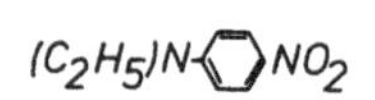	π^*	[260]
III	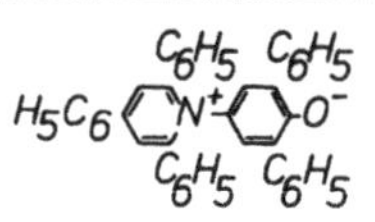	E_T	[8]
IV	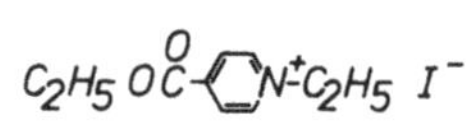	Z	[6]
V	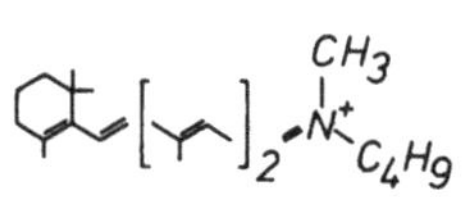	ΔE	[261]

of the pyridinium iodide complex (compound IV) are the basis of the E_T [8] and Z [6] scales. All these polarity scales represent the ability of solvents to solvate negative charges. An attempt was also made to classify the nucleophilic behaviour of solvents by the shift of the absorption band of N-retinylidenemethyl-n-butylammonium iodide [261] (compound V).

Molecular vibrations in solvent molecules and multiatomic ions, as investigated by IR and Raman spectroscopy [262], are sensitive to changes in the surroundings of the molecule or ion. The measureable effects are band shifts, removal of degeneracy, variation of band intensity and shape, and the occurence and disappearance of bands. The resolution of the bands which are often overlapping requires highly developed computer-aided techniques [263-268]. The measured effects are not very sensitive to concentration; so far only concentrated solutions, 0.1 M to saturation, have been studied. The use of Fourier-transform IR spectrometers and multiplexing techniques [269-271] will surely permit the use of lower concentrations and improve the accuracy of data.

IR and Raman-active intramolecular motions of ions or solvent molecules, giving evidence of solute-solute interactions, have been reported for many electrolytes in appropriate solvents, e.g. acetonitrile [272-280], dimethylsulfoxide [272, 281-283], liquid ammonia [284-289], methanol [290], formamide [291, 292].

Both IR and NMR data indicate that in methanol solutions solvation of both cations and anions occurs by primary solvent molecules which are strongly hydrogen-bonded to those in the bulk solvent [290]. This result has been confirmed and extended to other protic solvents through microwave investigations on alkali metal halides and salts of divalent cations in methanol, ethanol and formamide [293, 294]. Alkali salt ion pairs in protic solvents show cation-anion distances, as found by calori-

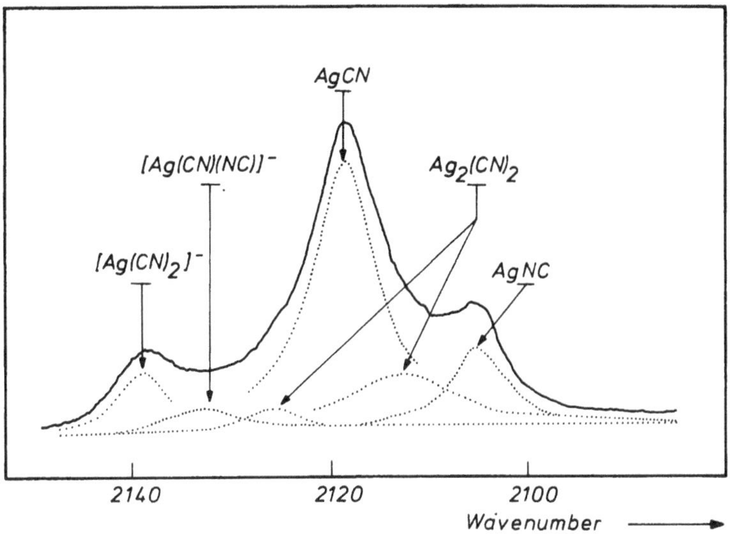

Fig. 10. Raman spectrum of silver cyanide in liquid ammonia and its decomposition into component bands. (with kind permission of the authors [289])

metric and conductance measurements, which require the inclusion of solvent molecules, and are in agreement with the spectroscopic evidence.

In his comprehensive investigation on Raman spectra of liquid ammonia solutions [288,289] Gill found multicomponent bands of multiatomic anions. The system AgCN/NH_3, Fig. 10, is given as an example. The positions and variations of relative intensities of component bands with changing cation, concentration and temperature indicate that a range of discrete ion pairs and higher aggregates coexist in equilibrium.

Association constants of electrolytes in organic solvents obtained from Raman and IR measurements are known, e.g. for NaI and KI in AN [275], $AgNO_3$ in AN [279], LiSCN in polar solvents and ethers [282,295,296], $NaCo(CO)_4$ in polar aprotic solvents [297,298] and $(C_4H_9)_4NCl$ in benzene [299]. The equilibrium concentrations of ions and ion pairs, needed for the calculation of association constants, are determined by integrating the appropriate bands and comparing the areas with reference areas which are related to known concentrations. This procedure furnishes 'apparent association constants' sometimes referred to without characterising them as such [275]. The drawbacks of this method of calculation are the uncertainty in the area determination and the lack of reliable activity coefficients in concentrated solutions.

The approximations of the theory for dilute electrolyte solutions cannot be used without ambiguity. In particular the assumption cannot be maintained that the activity coefficient of the uncharged ion pairs equals one. Irish [300] suggested a way to account for the activity coefficient of the ion pair by assuming its behaviour to be similar to that of a structured dipole molecule, e.g. glycine. The quantitative comparison of the spectroscopically-determined association constants with those determined by conductivity, which is seldom found in the literature, also suffers from the fact that in vibrational spectroscopy only short-range interactions can be observed. As a consequence of the last statement Janz et al. pointed out [279] that in acetonitrile $AgNO_3$ forms only contact ion pairs; they draw the conclusion from the rather satisfactory agreement between K_A(Raman) and $K_A(\Lambda)$.

The influence of cation-complexing agents like crown ethers [281], cryptands [283, 298] or biological macrocyclic systems [301] on the structure of ion pairs is very marked and can easily be seen in vibrational spectroscopy.

Most spectroscopic investigations have been made with polyvalent electrolytes which show marked association and solvation effects but for which almost no reliable information exists from conductance measurements.

7.4 NMR and Related Methods

Nuclear magnetic resonance is a sensitive probe of the immediate chemical environment of ions in solution [302, 303]. The use of the Fourier transform technique combined with high resolution spectrometers permits measurements on nuclei with spins $I \geqq {}^1/_2$ down to 10^{-3} M solutions [304]. Besides the classical nuclei with $I = {}^1/_2$ (^1H, ^{13}C, ^{15}N, ^{19}F, ^{31}P, ^{205}Tl etc.) there exist NMR-active isotopes with $I \geqq 1$ of almost all elements of the periodic system [305] (^2H, ^7Li, ^{11}B, ^{14}N, ^{23}Na, ^{25}Mg, ^{27}Al, ^{35}Cl, ^{39}K, ^{43}Ca, ^{81}Ba, ^{87}Rb, ^{133}Cs etc.). For nuclei with $I = {}^1/_2$ the position and the multiplicity of the chemical shift, δ, is due to spin-spin and spin-lattice coupling and to magnetic shielding. Nuclear spins of $I \geqq 1$ yield electrical quadrupole moments which interact with the electrical field gradient resulting from the charge distribution of the adjacent molecules. In this case an electrical field of non-spherical symmetry is required but this is caused even by thermal fluctuations. Magnetic dipoles give sharp signals but quadrupoles show band broadening.

The maximum value of the chemical shift increases with increasing atomic number. A 2000-fold augmentation of δ is observed when going from ^1H to ^{205}Tl, making thallium salts very sensitive probes for NMR investigations [306, 307]. ^1H-NMR spectroscopy is used mainly for investigation on hydrogen-bonding systems in aqueous and alcoholic solutions [290, 300]. The interaction between acceptor and hydroxyl group was examined in systems which are highly diluted with an inert solvent [308]. The chemical shift of ^{31}P of $(C_2H_5)_3PO$ is the basis of the acceptor numbers, AN [11].

Popov's comprehensive multinuclear NMR studies of alkali ions in non-aqueous solvents [304, 309–316], encompassing concentration-depèndence of chemical shifts, ion-pair formation, influence of the solvent and correlation with donor numbers, and the role of macrocyclic polyethers and cryptands, are evidence of the powerful tool provided by NMR methods for the investigation of non-aqueous electrolyte solutions.

With respect to the exchange-time of ionic processes, the frequency of the NMR signal is generally low. Consequently, only a population-averaged resonance signal indicates the equilibrium between free ions and ion pairs. The concentration-dependent chemical shift can be expressed by

$$\delta_{obs} = x'_{FI}\delta_{FI} + x'_{IP}\delta_{IP} \,. \tag{44a}$$

The mole fractions of free ions, x'_{FI}, and ion pairs, x'_{IP}, are linked as usual to the association constant, K_A [311],

$$K_A = \frac{c - c'_{FI}}{(c'_{FI})^2 \cdot y'^2_{\pm}}; \qquad x'_{FI} = \frac{c'_{FI}}{c}. \tag{44b, c}$$

δ_{FI} can be obtained by extrapolation to infinite dilution, K_A and δ_{IP} by nonlinear least-squares fitting [311]. The attempt [280, 317] to correlate association constants from NMR measurements with those deduced from electrical conductance is not very meaningful because the two methods look at different populations of the ionic surroundings.

Macrocyclic complexing agents produce chemical shifts of the complexed cations, which are independent of the solvent. The occurence of distinct signals demonstrates the stability of such complexes [318].

The determination of spin-lattice relaxation [319–321] gives the possibility of investigating dynamic properties of the electrolyte and the solvent in the solution. Attempts have been made to measure diffusion coefficients [322] and transport numbers [323].

7.5 ESR-Spectroscopy

ESR-spectroscopy of radical anions and cations uses the coupling of electron and nuclear spin as the probe for ion-ion and ion-solvent interactions. Information is obtained from analyzing band shifts and band broadening and determining g-values in various solvents [324].

7.6 Relaxation Methods

The frequency range of microwave (MW) and far infrared (FIR) investigations extends between $5 \cdot 10^8$ and 10^{13} Hz (0.01 to 200 cm^{-1}). Different names are used as a consequence of different measuring techniques [294] but all methods study the response of the solution to changing electromagnetic fields.

The microwave response both of polar solvents and electrolyte solutions is usually represented with the help of its frequency-dependent complex relative permittivity, $\varepsilon(\omega) = \varepsilon'(\omega) + j\varepsilon''(\omega)$, cf. Ref. [325]. The characteristic parameters of such investigations are the relaxation times or relaxation time distributions of molecular processes and the extrapolated 'real' permittivities of zero (ε_0) and inifinite (ε_∞) frequencies of one or more relaxation regions.

Figure 11 shows a representation of $\varepsilon''(\omega) = f(\varepsilon'(\omega))$, called an Argand diagram, for 0.48 M $NaClO_4$ in a PC-DME mixture (20 weight% PC). Data analysis of the precedingly determined frequency-dependent permittivities of the solvent mixture without $NaClO_4$ yielded two relaxation regions, one attributable to DME (relaxation time $\tau = 4.7$ ps) the other to PC (relaxation time $\tau = 22$ ps). The shifts of solvent relaxation times with reference to those of the two pure solvents, $\tau(\text{DME}) = 3.6$ ps and $\tau(\text{PC}) = 39$ ps, is correlated to the change in viscosity. Addition of the

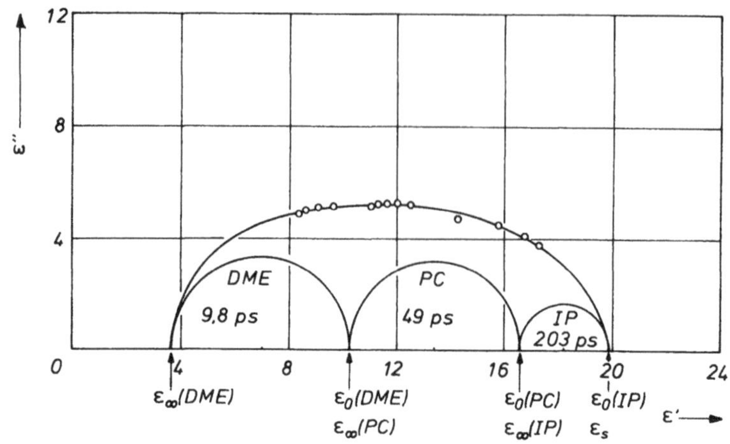

Fig. 11. Argand diagram ε'' vs. ε' of a 0.48 M $NaClO_4$ solution in propylene carbonate-dimethoxyethane (20 weight % of PC) at 25 °C (upper curve with measured points) [326]. Data analysis yields three relaxation times and appropriate Debye circles

electrolyte, Fig. 11, yields a third relaxation region (relaxation time $\tau(IP) = 203$ ps) due to ion-pair movements and shifting of the solvent relaxation times to higher values, $\tau(DME) = 9.8$ ps and $\tau(PC) = 49$ ps [326]. The static permittivity of the solution, ε_0 increases with increasing salt concentration [326].

Separate relaxation regions attributable to ion pairs [294, 326, 327] and higher aggregates [328] cannot be observed in all solvents. Free ions contribute to $\varepsilon''(\omega)$ through their conductances. A survey is given in Ref. [36]. As a rule, 1,1-electrolytes in protic solvents or solvents of high permittivity cannot be identified by separate regions. They yield decreasing ε_0-values (static permittivity of the solution) [293, 329, 330] with increasing electrolyte concentration which depend specifically on the ions, e.g. Li > Na > K > Rb > Cs in water and alcohol solutions. Figure 12 gives a survey on $LiNO_3$ solutions in various solvents [293].

Both broadening and shifting of the relaxation time distribution of these solutions are also ion-specific [329–331]. E.g. in methanol as solvent $NaClO_4$ shifts relaxation times to higher, Bu_4NClO_4 to lower frequencies indicating 'structure making' or 'breaking' by the added salt [331], cf. Fig. 13.

The FIR response (0.1 cm^{-1} to 100 cm^{-1}) of polar dielectrics is related to inertial effects and the libration of dipoles [294]. Ions in polar media contribute in a complex manner. Information is provided by Lambert-Beer's law related to the complex relative permittivity containing the conductance contribution to ε'' when electrolyte solutions are investigated. The rotational motion of polar molecules gives rise to a broad-band absorption with a maximum in the FIR region [332]. Models proposing the libration of each molecule in the cage of its neighbours are used to explain the excess absorption observed [294, 332, 333]. Ions in the solution interact specifically. For alkali salts a typical cation-band (about 400 cm^{-1} for Li^+, 110 cm^{-1} for Cs^+) is observed arising from cation-solvent vibrations [334–336]. The non-visible contribution of the anion mass excludes ion-pair vibrations, in contrast to tetraalkylammonium

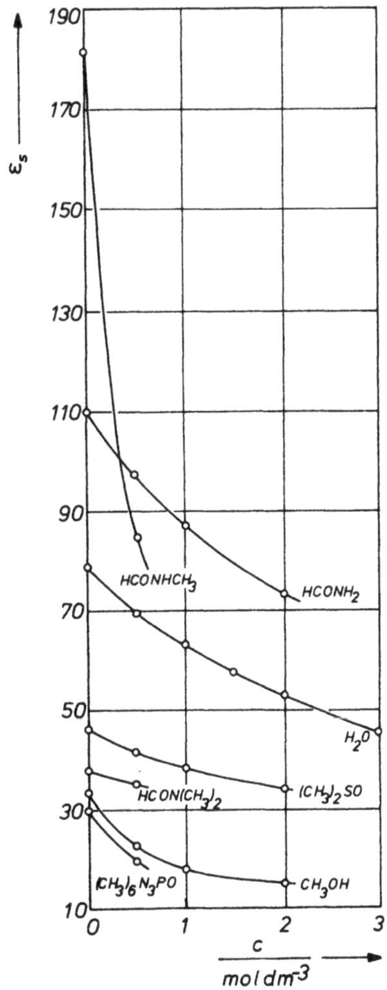

Fig. 12. Decrease of permittivity of various solvents by addition of $LiNO_3$ at 25 °C [293]

salts in solvents of low permittivity where a broad, strong and asymmetrical absorption band, e.g. at 80 cm^{-1} for Bu_4NBr, is attributed to the libration of ion-pair dipoles [294]. Lestrade, Badiali and Cachet have stressed the compatibility of FIR results and the chemical model [294].

The region of radiowaves and still lower frequencies provides information on the relaxation of the ion cloud [337-339] in electrolyte solutions.

In connexion with dielectric and other spectroscopic relaxation methods, e.g. NMR, the group of ultrasonic relaxation, temperature- and pressure-jump methods [51, 340-342] must be mentioned. These yield information on the processes in electrolyte solution and confirm the basic chemical model of free ions and ion pairs in the solution [342].

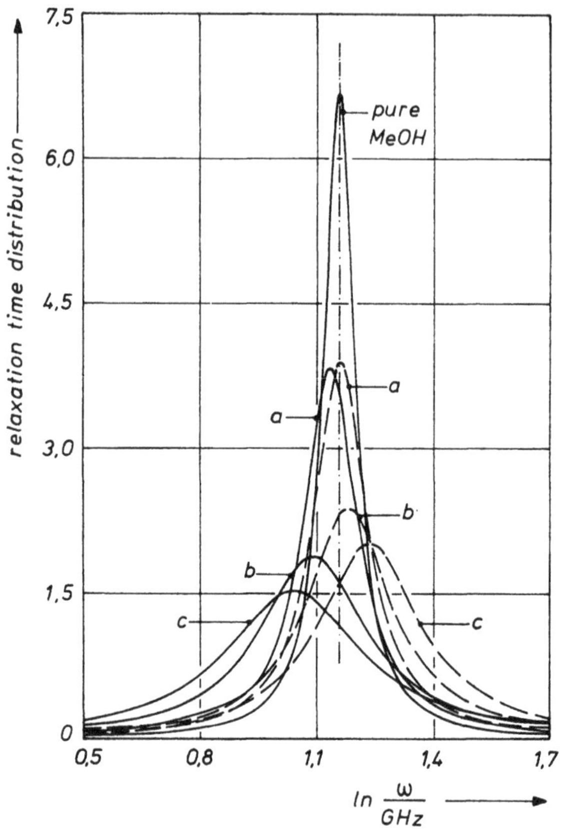

Fig. 13. Structure making and breaking effects in methanol solutions (25 °C) by tetrabutylammonium (full lines) and sodium (broken lines) iodide at concentrations 0.064 M (a); 0.250 M (b); and 0.535 M (c). This figure shows the shift and broadening of relaxation time distributions. For further details reference is made to Ref. [329]

VIII Chemical Reactions in Organic Solvents

The rate constants of chemical reactions [343], the yield [344] and the selectivity [345] of a reaction, as well as the conditions for refining or recycling [346] of products can be optimized by the choice of appropriate solvents. Discussion in this section is restricted to reaction mechanisms involving electrolytes or single ions. The role of electrolyte solutions in primary and secondary kinetic salt effects is not considered. For this problem see Refs. [78, 347–349].

8.1 Ions and Ion Pairs in Chemical Reactions

The ions and ion pairs of an electrolyte compound $Y = C^+A^-$, which is involved in a chemical reaction yielding an overall rate constant k_{obs}

$$B + Y \xrightarrow{k_{obs}} \text{Products} \tag{45}$$

may react by different reaction paths to yield the same reaction products [350–357], e.g.

$$B + \begin{cases} C^+ \text{ or } A^- \xrightarrow{k_{FI}} \\ [C^+A^-]_i^0 \xrightarrow{k_{IP}^{(i)}} \end{cases} \longrightarrow \text{Products} . \tag{46}$$

The ion pairs $[C^+A^-]_i^0$ are those which were introduced through Eq. (39), k_{FI} and $k_{IP}^{(i)}$ are the appropriate individual rate constants. Then the overall rate constant is given by the relationship

$$k_{obs} = \frac{k_{FI}c'_{FI} + \sum_{i=1}^{n} k_{IP}^{(i)}c'^{(i)}_{IP}}{c_Y}; \qquad c_Y = c'_{FI} + \sum_{i=1}^{n} c'^{(i)}_{IP} \tag{47}$$

and is strongly dependent upon the electrolyte concentration, c_Y. A simple case is given when data analysis can be executed with the help of an empirical set-up of equations of type of Eqs. (21)

$$k_{obs} = \alpha k_{FI} + (1 - \alpha)\, k_{IP} \tag{48a}$$

$$K_A = \frac{1-\alpha}{\alpha^2 c} \frac{1}{y'^2_\pm}; \qquad y'_\pm = \exp\left[-\frac{\varkappa q}{1 + \varkappa R}\right]. \tag{48b, c}$$

Figure 14 shows an example in which k_{FI} and k_{IP} are obtained from linear functions of the type $k_{obs}/(1 - \alpha)$ vs. $\alpha/(1 - \alpha)$.

Values of K_A from conductance measurements can often be used for calculating the degree of dissociation. Significant deviations, however, may also occur in this case, especially in mixed solvents, indicating that not all the pair configurations which are counted as ion pairs by the thermodynamic overall association constant, K_A, are reacting species [177].

Depending on the reaction mechanism various cases are observed, e.g.

(i) $k_{FI} > k_{IP}$ for the solvolysis of alkyl halides in ethanol [350,353].
(ii) $k_{FI} < k_{IP}$ cf. Fig. 14 [354].
(iii) $k_{IP} \sim 0$ for the reaction of alkali phenoxides with methyl iodide in alcoholic solutions [352],
(iv) $k_{FI} \sim 0$ for cyclic condensation reactions (Dieckmann condensation) [351].

8.2 Kinetic Solvent Effects

Comparison of reaction rates k^s in a solvent S and k^o in a reference solvent (index o) yields

$$\ln \frac{k^s}{k^o} = -\frac{1}{RT}\left[\Delta G_c^{\ddagger s} - \Delta G_c^{\ddagger o}\right] \tag{49}$$

or related relationships, e.g. Ref. [358,359].

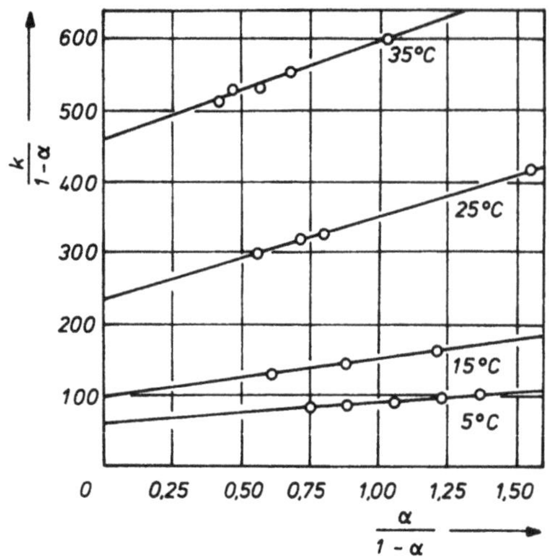

Fig. 14. Reactivity of ions and ion pairs in the reesterification reaction

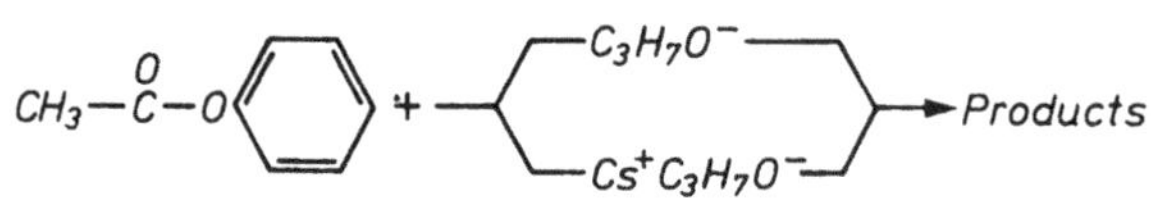

at various temperatures in propanol as the solvent [354]. Data analysis was executed with the help of Eqs. (23). Association constants of C_3H_7OCs/PrOH were determined by conductance measurements.

Temperature (°C)	5	15	25	35
$k_{FI}/(dm^3\ mol^{-1}\ s^{-1})$	0.52	0.88	1.78	2.47
$k_{IP}/(dm^3\ mol^{-1}\ s^{-1})$	1.07	2.12	4.0	7.68

The energies ΔG^* based on the particle densities of the reacting species represent the maximum reversible work to build the activated complex X^*, from the initial products in the reference solvent and in the solvent S, respectively. Eq. (49) is obtained from the theory of absolute reaction rates by assuming that the reaction mechanism is not affected by transferring the reaction from one solvent into the other. Change in the reaction mechanism is reported when the transfer is followed by a change of the charge distribution of the activated complex, e.g. [360].

Studies of substitution, addition-elimination, elimination, and addition reactions involving ions or charge-bearing activated complexes are the usual means of investigation of solvent effects. Table VIII gives a survey of these reactions.

Generally valid equations for a quantitative discussion of solvent effects cannot normally be obtained. The use of models in kinetic investigations is based on presumptive evidence. Experimental rate constants or activation energies are compared with the prediction of possible configurations of the activated complex. For this purpose models containing detailed charge distributions for calculating activation energies, ΔG^*, and their dependence on solvent parameters are available from the

Table VIII Appropriate classes of reactions for the investigation of solvent effects.
Y^{z_Y}: nucleophile of charge $-1 \leqq z_Y \leqq 0$
E^{z_E}: electrophile of charge $0 \leqq z_E \leqq +1$
R: general organic group; X: leaving group;
L: nucleophilic ligand;
Transition state formulae are in brackets

Reaction and reaction type		Examples and References	
I. Substitution			
I.1. $R\text{–}X \rightarrow [\overset{\delta+}{R}\cdots\overset{\delta-}{X}] \rightarrow$ Products	S_N1	$(CH_3)_3CCl \rightarrow (CH_3)_3COH$	[361]
		$(C_2H_5)_3S^+Br^- \rightarrow (C_2H_5)_2S + C_2H_5Br$	[362]
I.2. $Y^{z_Y} + R\text{–}X \rightarrow [Y\cdots R\cdots X]^{z_Y} \rightarrow$ Products	S_N2	Menschutkin reaction	[363][364]
		$RX + CN^- \rightarrow RCN + X^-$	[365]
		$(CH_3)_3N + (CH_3)_3S^+ \rightarrow (CH_3)_4N^+ + (CH_3)_2S$	[366]
		$(CH_3)_3S^+ + OH^- \rightarrow (CH_3)_2S + CH_3OH$	[367]
		$L + [MeS_n] \rightarrow [LMeS_{n-1}] + S$	[368]
II. Addition - Elimination			
$Y^{z_Y} + R\text{–}\underset{X}{C}{=}Z \rightarrow [Y\cdots \overset{R}{\underset{Z}{C}}\cdots X]^{z_Y} \rightarrow$ Products	$A_{Ac}2$	acid ester solvolysis	[369]
	$B_{Ac}2$	neutral ester solvolysis	[370]
		alkaline ester solvolysis	[371]
III. Elimination			
III.1. $H\text{–}C\text{–}C\text{–}X \rightarrow [H\text{–}C\text{–}C\cdots X] \rightarrow$ Products	E_1	β-Eliminations	[372]
III.2. $Y^{z_Y} + H\text{–}C\text{–}C\text{–}X \rightarrow [Y\cdots H\cdots C\text{–}C\cdots X]^{z_Y} \rightarrow$ Products	E_2	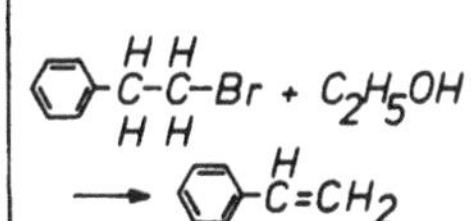	[373]
IV. Addition			
$E^{z_E} + {>}C{=}C{<} \rightarrow [{>}\underset{E}{C\text{–}C}{<}]^{z_E} \rightarrow$ Products		${>}C{=}C{<} + Br_2 \rightarrow \underset{Br}{\text{–}C}\text{–}\underset{Br}{C\text{–}}$	[374][375]

literature [78,376–280]. Appendices B and C contain the fundamentals of the extended chemical model (see Sect. IV) for these calculations.

The Gibbs energy of activation, $\Delta G^{\neq}$, can be related to the mean-force potential of the pair distribution function W_{ij} of the reactants which is written for this purpose in the following form

$$W_{ij} = [e_0 z_j \psi_i(P) + \mu_j \nabla\psi_i(P) + \theta_j \nabla\nabla\psi_i(P)] + W^*(P) \tag{50}$$

where e_0z_j, μ_j and θ_j are the charge, dipole moment and quadrupole tensor of particle j, $\Psi_i(P) = \psi_i(r, \theta, \varphi)$ the potential of the electric field of particle i at point P, and W* a possible specific contribution from solvent-particle interactions.

Only a few simple examples of applications will be given. For ion-ion reactions the solvent effect, Eq. (49), is controlled by the relationship

$$\ln \frac{k^s}{k^0} = \frac{e_0^2 z_i z_j}{4\pi\varepsilon_0 kT} \frac{1}{r_0} \left[\frac{1}{\varepsilon^0} - \frac{1}{\varepsilon^s}\right] + \frac{W^{*0} - W^{*s}}{kT} \tag{51}$$

and for ion-dipole molecule reactions by

$$\ln \frac{k^s}{k^0} = \frac{e_0 z_j}{4\pi\varepsilon_0 kT} \frac{6(\varepsilon^s - \varepsilon^0)}{(2\varepsilon^0 + 1)(2\varepsilon^s + 1)} \frac{\mu_i}{r_0^2} \cos \vartheta_0 + \frac{W^{*0} - W^{*s}}{kT}. \tag{52}$$

Both equations are obtained from the general equations, appendix B when the reacting ions are represented as single charges and the reacting dipole-molecules as point-dipoles in the spheres of radius a. Quantities r_0 (Eq. (51)) and r_0, ϑ_0 (Eq. (52)) are the coordinates of particle j which characterize the configuration of the activated complex in the coordinate system of particle i.

By careful selection of a series of solvents or solvent mixtures the term due to specific interactions, $W^{*0} - W^{*s}$, may be eliminated and then the well-known equations of the literature [376-380] are obtained. At an early stage of the research on solvent effects, Hughes and Ingold [381] established useful qualitative rules for nucleophilic substitution and addition-elimination reactions based on the permittivity of the solvent. Examples of changing the permittivity in the vicinity of reactants as a consequence of hydrophobic and hydrophilic interactions were given recently [382]. The important variation of the rate constant for the addition reaction of halogen to olefines [374,375] in polar solvents is a further example of interest. A good conformity between the chemical model and the experiment is obtained when an interaction of a halide cation and a non-polar but polarizable molecule (olefine) can be assumed [376]. The basic equations for calculating interactions between ions and polarizable molecules are given in Appendix C. The discussion of solvent effects in the framework of these models has been limited to series of solvents with similar structures or to solvent mixtures where one of the solvent components changes the permittivity of the bulk solvent but is inert in the solvation of the reactants and the activated complex.

The suppression of the interaction term, $W^{*0} - W^{*s}$, means the neglect of changing specific solvation when changing the solvent. The influence of solvents belonging to different solvent classes (Table I) which change the rate constant of some reactions by many powers of ten, reveals the contribution of short-range, specific solvation predominating the Coulombic interactions.

Medium activity coefficients, cf. Section 5.5, can be used to discuss these effects. Fig. 15 provides a summary of the changes in solvation energy of ions and neutral molecules of various types in solvents which are representative for the solvent classes of Table I. The energy scale, $RT\ln_{MeOH}\gamma_s$, with methanol as the reference solvent is taken from Ref. [359]. The non-measurable medium activity coefficient of the activated complex can be estimated from similar stable molecules or ions.

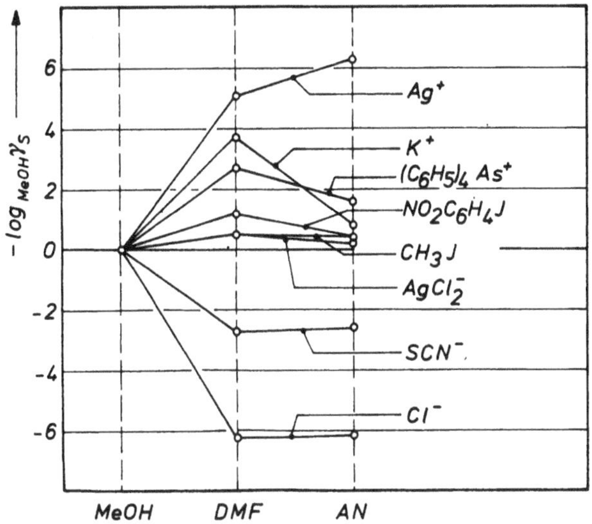

Fig. 15. Solvent activity coefficients for various ions and molecules with respect to MeOH as the reference solvent

In a comprehensive investigation Parker [359)] tested the consistency of the model calculation through the conformity between the estimated values and the experimental values from kinetic measurements. The extra-thermodynamic equi-partition assumption, e.g. using $(AsPh_4)^+$ $(BPh_4)^-$ as the reference electrolyte, was adopted for all particles, thus vanishing in the final equations.

Table IX shows six examples of reactions in almost isodielectric solvents. The increase of up to six orders of magnitude in the rate constant of the reactions 1) to 3) is the result of drastically decreasing anion solvation and weakly increasing solvation of the activated complex in the dipolar aprotic solvents with a minimum effect for the soft anion, SCN^-, which is better solvated in AN or DMSO than hard anions. The stronger solvation of the polar activated complex in dipolar aprotic solvents and

Table IX Solvent effect ln (k^s/k^0) in isodielectric solvents. k^0: rate constant in methanol as the reference solvent

solvent	MeOH	DMF	AN
permittivity (25 °C)	32.7	36.7	35.9
solvent class (Table I)	1	4	5
Reaction	log (k^s/k^0)		
1) $CH_3I + Cl^-$	0	5.9	4.6
2) $CH_3I + SCN^-$	0	2.2	1.4
3) p-$NO_2(C_6H_5)I + N_3^-$	0	4.5	3.9
4) $(CH_3)_3S^+ + N_3^-$	0	3.1	3.7
5) n-BuCl + Pyridine	0	−0.2	—
6) $(CH_3)_3S^+ + (CH_3)_3N$	0	0.8 (DMA)	—

competition between decreasing anion and increasing cation ion-solvent interactions are observed in reaction 4). Reactions between molecules forming an uncharged activated complex, reaction 5), or reactions in which the charged species are a cation and an activated complex of equal charge, reaction 6), are only weakly affected by the transfer from polar protic to aprotic solvents. The solvation of the reactants in reaction 5) increases in DMF and AN, rendering the reaction slower in aprotic media. For further discussion of specific solvent effects see Refs. [383,384].

It is not possible to transfer hydroxyl ions into dry aprotic dipolar solvents. Therefore the alkaline hydrolysis of esters, which is one of the most investigated second-order model reactions in water and aqueous mixtures, is unsuited for the investigation of specific solvent effects despite many efforts [385-388].

Most inorganic reactions of solvolysis and nucleophilic substitution of transition metal complexes are rather insensitive towards a change from protic to aprotic dipolar solvents [368]. Since the maximum coordination number of the complexes cannot be exceeded, in most cases the rate-determining step is a dissociation leaving a transition state with a positive charge almost one unit higher than the initial complex and an anion or, in few cases, a molecule. Thus the decreasing solvation of anions by dipolar solvents is counteracted by the increasing solvation of the cationic transition state.

The solvent effect on proton-transfer reactions is determined by two effects [389]. In cases of slow proton transfers, which are not diffusion-controlled, e.g. reactions between carbon acids and weak bases [390], the solvent effect is determined by hydrogen-bonding and formation of ion pairs whereas the viscosity of the solvent prevails in very fast diffusion-controlled reactions.

8.3 The Use of Correlation Functions

The tight link between kinetic solvent effects and the theory of solvation, suggests the correlation of kinetic data with parameters quoted in Section II when quantitative information from the theory of solvation is not available.

Nucleophilic effects (solvation of Lewis acids) can be correlated with Gutmann's DN-values [11] or with Palm's B-coefficient [391]. The two coefficients are positively correlated but show marked deviations for hydrogen bonding systems. Electrophilic effects (solvation of Lewis bases) are represented by the E_T-values of Dimroth and Reichardt [8], Kosower's Z-scale [6,7] or the AN-number of Gutmann and Mayer [15]. The E_T and Z parameters show a strong positive correlation which can be understood from the similar basis of both scales (cf. Sect. VII). The correlation between E_T and AN-numbers is also positive [15]. Inasmuch as these parameters are based on enthalpy effects the simultaneous correlation of protic and aprotic solvent properties yields usually two functions — one for protic and the other one for aprotic solvents — when entropy is the determining factor, as in hydrogen bonding systems.

The correlation of kinetic data with empirical parameters is based on the principle of Linear Free Energy Relationships (LFER) and Quantitative Analogy Models (QAM) [392,393].

In a general way, correlation functions link a measurable quantity Z_k of a system, the i-th influence of the environment, ξ_{ik}, and the response of the system toward this influence, β_i, by a linear expression

$$Z_k = Z_0 + \sum_i \beta_i \xi_{ik} \tag{53}$$

which is the result of Taylor-series linearisation of the quantity Z_k. Z_0 is the quantity Z_k under normalized conditions. Today this method is widely used in computer-assisted research.

Application to kinetic solvent effects consists in choosing $\ln k_s$ as the property Z_k and the above quoted correlation parameters as the quantities ξ_{ik}.

In the simplest case Eq. (53) can be represented by a one-parameter correlation

$$\ln k^s = \alpha + \beta\xi_s , \tag{54}$$

where $\alpha = \ln k^0$ and $\xi_s = DN$, E_T, or AN etc. In many cases such one-parameter correlations yield satisfactory linear functions when solvents of the same class or related classes are considered [12,394–398].

Multiparameter correlations of the type

$$\ln k^s = \alpha + \sum_i \beta_i \xi_{is} \tag{55}$$

are of increasing interest because of the improved computer techniques of multiple correlation.

Koppel and Palm used an equation with four parameters [391].

$$\ln k^s = \ln k^0 + eE + bB + pP + yY . \tag{56}$$

The equations of Fawcett and Krygowski [399,400]

$$\ln k^s = \alpha + \beta_1 E_T + \beta_2 DN \tag{57}$$

and of Mayer [401]

$$\ln k^s = \alpha + \beta_1 \Delta DN + \beta_2 \Delta AN - \beta_3 \frac{\Delta[\Delta_{vap}G]}{RT} \tag{58}$$

yield good correlation encompassing many solvent classes.

Taft, Kamlet and coworkers [261] separate the solvent polarity from hydrogen bond donor (HBD) and hydrogen bond acceptor (HBA) effects by the relationship

$$\ln k^s = \ln k^0 + n\pi^* + a\alpha + b\beta \tag{59}$$

π^* is the frequency shift of p-nitrophenyl dyes, for which hydrogen bonding is excluded, and α and β are HBD and HBA-properties of the solvent respectively. The polarity scale π^* is correlated with the dipole moment of the solvent, corrected with respect to the average reaction field [402]. The HBA-coefficient β is correlated with the proton affinity in the gas phase [403].

For nonpolar solutes, e.g. n-octane, it has been shown that the Gibb's energy of solution in different solvents is linearly correlated with Hildebrand's solubility parameter δ_H [404], which takes into account the van der Waals and dispersion forces. In Eq. (58) these forces are considered by the change of the Gibbs energy of evaporation of the solvent.

A new generation of multiparameter correlations of high technological importance is presented in the comprehensive work of Dubois on the DARC system [405,406] and of Fredenslund and Rasmussen [407,408] and others on the UNIFAC method. These permit the advanced calculation of reactivity and solution properties on the basis of topological treatments of the molecules and their vicinity from a minimum of experimentally-determined supporting data. These methods form the basis of comprehensive data banks [409–412].

8.4 Phase-Transfer Reactions

The principle of phase-transfer reactions — advantageously used for substitution, alkylation, acylation, elimination, and redox reactions of organic compounds, RQ, — is the transfer of an ion from aqueous solution into an apolar phase where the chemical reaction

$$[C^+A^-]^o_{org} + RQ \rightarrow RA + [C^+Q^-]^o_{org} \tag{60}$$

takes place [413–416]. Two features previously shown for solvent effects are thus used, the very high reaction rate of nucleophilic reactions in aprotic solvents and the reactivity of ions and ion pairs in chemical reactions. The problems of water transfer are minimized by the choice of apolar solvents, immiscible with water and suitable lipophilic counterions of the reacting ion.

The ionic co-reactant is usually an anion A^-(Cl^-, N_3^-, OH^-, CN^-, NO_2^-, F^-, RCO_2^-, RO^-, BH_4^-, MnO_4^-, ClO_4^-) which is transferred into the apolar phase (benzene, toluene, pentane, CH_2Cl_2, $CHCl_3$, CCl_4) by the help of a 'catalytic' cation C^+(R_4N^+, R_4P^+, R_4As^+, metal cation complexed with crown ethers or cryptates) suitable for forming an ion pair, $[C^+A^-]^0$, in the apolar solvent.

The reaction, Eq. (60), is controlled by the heterogeneous equilibrium

$$A^-_{aq} + [C^+Q^-]^o_{org} \rightleftarrows Q^-_{aq} + [C^+A^-]^o_{org} \tag{61a}$$

with the equilibrium constant

$$K = \frac{[C^+A^-]^0_{org} \times [Q^-]_{aq}}{[C^+Q^-]^0_{org} \times [A^-]_{aq}} \,. \tag{61b}$$

Thus the ion-pair concentration $[C^+A^-]^o_{org}$ can be controlled *via* the concentration of A^- in the aqueous phase. The appropriate choice of electrolyte concentration, solvent and the catalyst are the conditions for optimizing the process. Tetraalkylammonium salts are commonly used and the number of carbon atoms should be at least 16. Unsymmetrical ions with one long alkyl chain tend to form micelles and should be avoided [417]. Appropriate catalysts are Bu_4N^+, TEBA® = $= N^+(C_6H_5CH_2)(C_2H_5)_3$, and ALIQUAT 336® which is a mixture of trialkylammonium salts with 15 to 30 carbon atoms; Bu_4NHSO_4 is very suitable because of its cheapness and the ease of recovering it from the organic phase by treatment with sulfuric acid. In the anion sequence $F^- > OH^- > HSO_4^- > Cl^- > CN^- > Br^- > I^- > ClO_4^-$ the exchange equilibrium of the phase-transfer catalyst decreases from left to right and hence also its catalytic activity.

Phase-transfer catalysts are used to extract organic or pharmaceutical products from waste water [418] and to transfer hydrophilic dyes from water into hydrophobic polymers [419].

A convincing example used in organic and pharmaceutical synthesis [418,420] is the alkylation of CH, OH or NH-acids by alkylating agents with NaOH under mild conditions. The potential Hoffmann degradation of the transfer-catalyst ion [421] when tetraalkylammonium salts are chosen can be avoided by the use of the more expensive alkali crown complexing agents. For comparison, in non-aqueous homogenous phase reactions the alkylation takes place only with strong bases, like $NaNH_2$, NaOR, $LiOC_4H_9$ etc.

The use of chiral catalysts introduces the possibility of synthesizing stereospecific products [422] with only minor formation of racemates.

Part B
Technical Applications and Applied Research

IX Introduction

A perusal of recent literature shows an increasing interest in technical applications and applied research based on non-aqueous electrolyte properties. The assortment of solvents with widely varying properties, an almost unlimited number of solvent mixtures and soluble electrolyte compounds provides flexibility in tackling a given problem. The unique properties of non-aqueous solutions can be the key in solving special technical problems.

Primary batteries (Sect. X) of high energy density, very low self-discharge (long shelf life) and good low-temperature performance make use of the large liquid range of organic solvents and of the kinetic stability of lithium metal in non-aqueous solutions. They are commercially available and have replaced conventional systems in some fields, e.g. cardiac pacemaker batteries and military applications. Good solubilities for some technically important electrolytes and many non-electrolytes, large liquid range, solution stability, enhanced stability of the solvent against oxidation and reduction are factors which recommend non-aqueous electrolyte solutions for use in the electrodeposition of metals (Sect. XIII), the production of wet capacitors (Sect. XIV), and various fields of electro-organic synthesis (Sect. XV). Flat non-emissive electrode displays (Sect. XI) and liquid junction solar cells (Sect. XII) are further fields which are developing fast and where non-aqueous electrolyte solutions are successfully competing with solid state ionics and aqueous electrolyte solutions. Here the non-aqueous solutions may be superior to aqueous solutions because of increased corrosion resistance of the electrodes. Other built-in advantages when compared to solid state electrolytes are the better levelling properties both for temperature and concentration discontinuities and the good permanent contacts between electrodes and ionic conductors which are not interrupted by temperature and pressure changes or volume changes occuring from chemical reactions (high energy batteries, electrochromic displays, liquid junction solar cells).

Drawbacks of non-aqueous electrolyte solutions include the appreciately higher costs of the solvent and, in some cases, its toxicity or flammability. Until now some promising fields of application have been blocked because of existing alternative production units and by the technological problems of scaling up the electrochemical processes. The flexibility in the choice of appropriate electrolyte solutions is linked to an increased number of variables and may involve time-consuming optimisation

Table X Conductivities of several technically important ion-conducting systems

System	Temperature $\frac{\theta}{°C}$	Conductivity $\frac{10^4 \varkappa}{\Omega^{-1}\,cm^{-1}}$	Ref.
*5.68 M HCl/H_2O	+25	8490	423)
*6.80 M	−20	3529	
2.81 M $LiClO_4/H_2O$	+25	1517	424)
2.84 M	0	878.4	
*3.92 M $LiClO_4$/MeOH	+25	498.1	425)
*2.68 M	−45	123.7	
*0.66 M $LiClO_4$/PC	+25	54.20	232)
*0.34 M	−45	2.75	
*1.39 M $LiClO_4$/PC, DME 42 weight % of PC	+25	145.9	227)
*0.74 M $LiClO_4$/PC, DME 28 weight % of PC	−45	33.1	
LiCl (liquid)	+637	58540	426)
LiCl/KCl (eutectic) 58.5 mol % of LiCl	+475	16150	426)
Ethylpyridinium bromide/ $AlCl_3$, 1:2, melt	+25	84.3	427)
$LiO_2C_2F_3$-polyethyleneoxide complex	+25	10^{-4}	428)
LiI (s)	+25	10^{-3}	429, 430)
LiI/Al_2O_3	+37	1	429, 431)
LiI/Al_2O_3	+300	50 to 100	429)
Li_3N	+25	10	432)

studies. When compared to aqueous electrolyte solutions or salt melts, the non-aqueous solutions show distinctly lower conductivities, but their accessible temperature range is more favourable for many applications. Table X gives a comparison of some ionic conductors which have been used technically. The asterisk (*) means that the concentration variables were chosen to yield the maximum conductivity of the system, $\varkappa_{max}$, cf. Fig. 7.

In various fields the use of non-aqueous electrolyte solutions has reached the stage of commercialisation and offers the possibility of introducing both new products and procedures. The following seven sections provide information on the current state of knowledge in some of them. Advantages of the non-aqueous electrolyte solutions are best demonstrated by quoting numerical values proving the degree of optimization. We also felt that it was necessary to demonstrate the reasons which led to the selection of particular non-aqueous electrolyte solutions by including a short description of the competing technological alternatives. Preference has been given to the most recent publications in our choice of examples. For the earlier literature, reference is generally made to monographs and reviews.

Technology considers electrochemical systems, *i.e.* heterogeneous systems with conducting liquid and solid phases, surface layers, membranes, etc. The fields of research which were discussed in the preceding eight sections concern almost exclusively the electrochemical properties of homogeneous liquid systems. Solvent structure, ion solvation, and association and their competition turn out to be the governing factors for the understanding of electrolyte solutions. In many cases an understanding of the role of the electrolyte solution in electrochemical systems also requires consideration of the neighbouring phases of the solution.

It is satisfactory to state that modern electrolyte theory can be helpful for optimizing forthcoming technologies. On the other hand, exciting new problems arise from technological investigations which then stimulate the theory to search for new approaches.

X High-Energy Batteries

10.1 Background

High energy batteries [229] work with distinctly higher (1.5 to 10 times) [433] energy densities and higher open-circuit and working voltages than conventional systems. They may be operated even at low temperatures, e.g. —50 °C. Furthermore, they exhibit appreciably lower self-discharge rates (down to < 1 %/a) which in the case of lithium batteries are mainly due to film formation on the Li-electrode. Lithium is the most favoured anode material because of its low equivalent weight but Na [434,435], K [436], Mg [229,437–439], Al [229,439–441], and Ca [438,442,820] have been considered as possible substitutes. The reasons are the natural abundance of aluminium, magnesium or calcium, their higher electronic conductivity [442], and the higher security (higher melting points) provided by these metals.

Batteries can be divided into two classes, the non-rechargeable primary and the rechargeable secondary batteries. Various types of small primary batteries have been commercially available since the early seventies. Secondary batteries are now the object of increased research. Some success has already been gained by the invention of reversible cathode materials which, in conjunction with lithium anodes and appropriate electrolyte solutions, allow an increased number of charge-discharge cycles.

High-energy batteries with a lithium anode are classified [443,444] with regard to the type of their ionic conductor. This can be a fast solid Li^+-ion conductor, a fused lithium salt, a lithium-potassium-salt eutectic mixture, or a non-aqueous lithium salt solution. If inorganic solvents are used, e.g. SO_2, $SOCl_2$, SO_2Cl_2, the solvent itself is the depolarizer and then a solid catalytic electrode is needed, e.g. carbon. The type of ionic conductor determines the internal resistance of the cell and the working temperature range and hence the possible technical applications.

Systems which work only at elevated temperatures, e.g. Na/β-alumina/S or Li/salt melt/FeS_n, have a potential for traction and load-levelling purposes. The reader is referred to reviews [426,434,445–451].

Solid-state cells for operation at ambient temperatures are mentioned for comparison with the wet cells. The low conductivities of fast solid ion conductors at ambient temperatures, cf. Li_3N and LiI (Table X) limit their use to fields where low discharge currents can be tolerated, e.g. batteries for cardiac pacemakers.

An example is the Li/LiI/I_2 · PVP (PVP = poly-2-vinylpyridine) cell [433,444,452,453] which is commercially available. The cell reaction $14\,Li + PVP \cdot 8\,I_2 \rightarrow 14\,LiI + PVP \cdot I_2$ [431] entails an increase of the internal resistance during discharge (specific resistances [431] $10^3\,\Omega\,cm$ (PVP · 8 I_2); $10^7\,\Omega\,cm$ (PVP · I_2)). Typical characteristics are: open-circuit voltage [453]: 2.8 V; energy density [453]: 200 W h/kg or 0.7 W h/cm^3; self-discharge [431,453] <10% in 10 years; capacity [431]: 2 A h; current drain [431]: 25 μA. About 500,000 cells [452] (Cardiac Pacemakers Inc.; Catalyst Research Corp.; Medtronic Inc.; Wilson Greatbach Ltd) were implanted in the period 1972–80. A

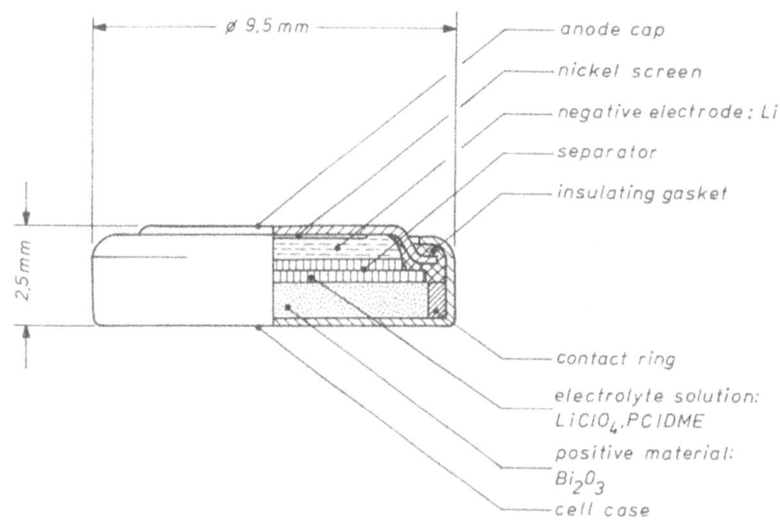

Fig. 16. Structure of a Li/Bi_2O_3 cell (VARTA) (with kind permission of VARTA)

promising new solid state-cell [428] uses $LiO_2C_2F_3$ in polyethylene oxide with a Li_xTiS_2 intercalation electrode (cf. also Sect. 10.5.2.). For further information see Refs. [431,454–456].

Lithium batteries with non-aqueous solutions are also available [443,444,453,457] commercially. Energy density, 300 W h/kg [444] or 0.35 to 0.5 W h/cm³ [458], and self-discharge, <2% per year [443,444] are comparable to those of commercial solid-state cells. The $Li/LiClO_4(PC)/Ag_2CrO_4$ cell has found extensive use in pacemakers, more than 350,000 being manufactured up to 1980 [452]. A diagram of a battery with a non-aqueous electrolyte solution is given in Fig. 16. Further information is given in Section 10.4.

Solid-state and wet lithium cells have replaced [453] the conventional Rubin-Mallory zinc/mercury pacemaker battery using aqueous NaOH solutions. Both types of lithium cell exhibit higher energy densities than conventional cells, e.g. Zn/HgO with 100 W h/kg [444] or 0.35 W h/cm³ [457] or Zn/MnO_2 with 0.1 to 0.2 W h/cm³ [458], and a far longer shelf-life, e.g. a lead-acid battery exhibits a self-discharge rate of 1% per day [444], the Zn/MnO_2-cell of 20% per year [444].

10.2 Non-Aqueous Electrolyte Solutions in Lithium Batteries

The requirements concerning non-aqueous battery electrolyte solutions are high specific conductance of the solution ($>5 \times 10^{-3}$ S cm^{-1}) and high mobility of the active ion over a large temperature range (−50 °C to +50 °C), sufficiently high solubility of the electrolyte compound (>0.3 mol dm^{-3}) at all temperatures, compatibility both with the lithium anode and the positive material (cathode) and, in the case of rechargeable batteries, stability over a sufficiently large voltage range.

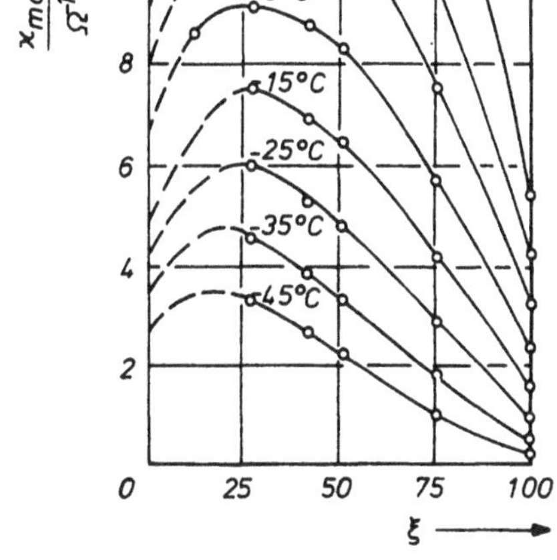

Fig. 17. Plot of maximum specific conductance of non-aqueous $LiClO_4$ solutions ($\varkappa_{max}$) vs. solvent composition (ξ) of the mixed solvent propylene carbonate-dimethoxyethane at temperatures (−45 °C ≦ θ ≦ +25 °C) [227]

Hence the optimum solvent for a battery would be a solvent of low viscosity, high permittivity, and low vapour pressure fulfilling all compatibility and stability conditions. Solvents of classes 6 or 7 in Appendix A-1, have low viscosities accompanied, however, by low permittivities and high vapour pressures, whereas those of classes 4 and 5 have high permittivities and low vapour pressures, but large temperature coefficients of viscosity yielding low conductances at low temperatures, cf. DME (class 6) and PC (class 5) in Figs. 17 and 18. Solvents of the other classes are unsuitable with regard to the stability and compatibility conditions. Solvent mixtures can improve both the solubility of the electrolyte and the conductivity of the solution and may reduce the temperature coefficient of conductivity and viscosity.

An example is $LiClO_4$ in PC/DME mixtures which is widely used as an electrolyte solution in commercial primary batteries. The complete information about the specific conductance of this battery electrolyte in the temperature range $-45\ °C \leqq \theta \leqq +25\ °C$ is given in Fig. 17 [227,232]. The viscosity and the relative permittivity of the mixed solvent over the same temperature range are shown in Figs. 18a [460] and 18b [326].

Figure 17 is based on series of conductance measurements on $LiClO_4$ in PC/DME mixtures as a function of solvent composition (weight % of PC = ξ), electrolyte concentration m, and temperature θ. For every solvent composition ξ, a $\varkappa$-m-function, see Fig. 7, was established by a least-squares procedure. The maximum specific conductances from these functions, $\varkappa_{max}$, are then plotted in Fig. 17 as a function of solvent composition and temperature. Figures 18a and b show the viscosity and the relative permittivity of the mixed solvent as a function of the same parameters. The conductance behaviour of the $LiClO_4$/PC/DME-system can be understood from the competition between the solvent viscosity, ion solvation and ion aggregation to ion pairs, triple ions and higher aggregates [233]. A comprehensive study of this

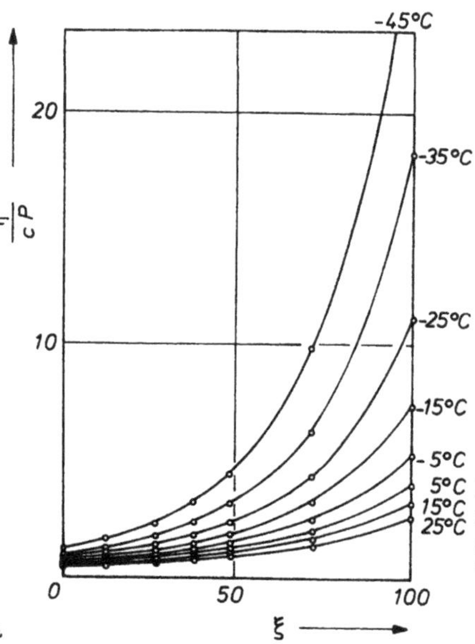

a

Fig. 18a. Viscosity of propylene carbonate-dimethoxyethane mixtures (ξ = weight % of PC) at various temperatures ($-45\ °C \leqq \theta \leqq 25\ °C$) [460]

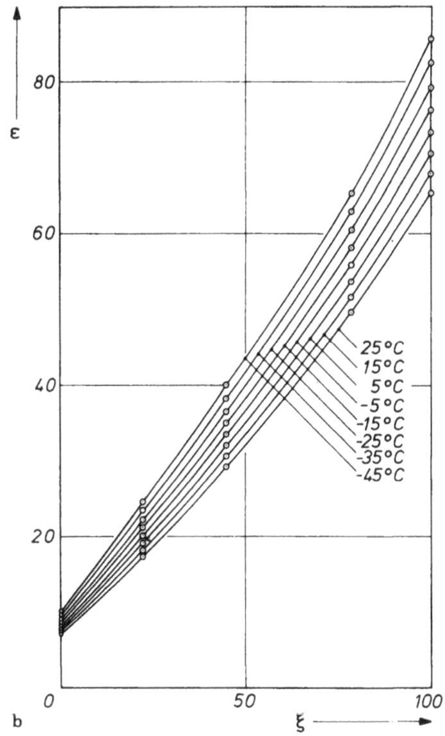

Fig. 18b. Relative permittivity of propylene carbonate-dimethoxyethane mixtures (ξ = weight % of PC) at various temperatures ($-45\ °C \leqq \theta \leqq 25\ °C$)[326]

type on 12 further electrolyte compounds in PC, DME and PC/DME mixtures from infinitely dilution to saturation provides useful rules for the choice of battery electrolytes, see Section 6.2.

Competition of solvation and association affects conductance strongly. Conductivity enhancements up to a factor of 300 were observed when DME, triglyme, DMSO, HMTT, PC, AN or 12-crown-4 ether were added to 1,3-dioxolane solutions of LiI, LiSCN and $LiBPh_4$ [461,463]. Increase of ion mobilities up to a factor of 5 is reported for additions of 12-crown-4 ether to THF solutions of $LiBF_4$ [464]. Margalit et al. [466] compared the conductance behaviour of $LiAsF_6$ in γ-BL/PC and DME/PC mixtures. Like us, [227,232], Venkatasetty et al. [467] stressed the possibility of rationalizing the discussion of concentrated electrolyte solutions in terms of dilute solutions, their example being $LiAsF_6$ in THF, NM, and dimethylsulfite. The use of multicomponent organic solvent mixtures is illustrated by the following examples: $LiBF_4$, $LiPF_6$, or $LiAsF_6$ in mixtures of 3-methyl-2-isoxazolidone/dioxolane/THF/EGS [468] and $LiClO_4$ in THF/PC/DME [469].

The relative mobility of the active ion, Li^+, in lithium batteries can be improved by the choice of appropriate counterions, (see transference numbers, Sect. 6.1.) Lithium salts with large organic anions like $Li[B(C_2H_5)_3\ (C_4H_4N)]$ [470], or halo-organic metal salt complexes like $Li[B(C_6F_5)_3CF_3] \cdot DEE$ [471] are suitable candidates. Such salts have the further advantages of high solubility, low ion association, and replacement of ClO_4^- which might be hazardous [472,473]. Further information may be found in Refs. [230,457,462,465,474–477].

10.3 Stability of Electrolyte Solutions with Lithium

The most important condition for the use of non-aqueous electrolyte solutions in lithium batteries is their stability in contact with lithium. At least kinetic stability is required. Of the many solvents which are claimed to be stable with lithium [438] few if any [478,479] seem to be thermodynamically stable.

Propylene carbonate, the solvent used most widely in commercial batteries, decomposes [480,481] cathodically on graphite at potentials of >0.6 V against a lithium reference electrode: $2\,Li^+ + PC + 2\,e^- \rightarrow CH_3{-}CH{=}CH_2 + Li_2CO_3$. This reaction is also found with lithium amalgam [482]. Kinetic stability is supposed to result from a passivating film on the substrate which protects the lithium from further reaction with PC. This film could be Li_2CO_3, cf. Refs. [480–483], or a polymerisation product of propylene, cf. Ref. [484]. Epelboin et al. [484] observed both Li_2CO_3 in the pores of the substrate and a plastic layer which they consider to be the passivating film. Despite the importance of these results, for batteries and electrodeposition (Sect. XIII), only a few solvents have been comprehensively investigated with regard to their reactivity. Further examples are AN [230,485,486], DMF [485], NB [485], and γ-BL [487], Caiola et al. pointed out rules [488,489] for the apparent stability of organic solvents with lithium. Cyclic molecules were found to have greater stability than open-chained ones; molecules with short alkyl chains are more stable than those with long ones.

Dey and Holmes reported [486,490] the reactivity of organic solvents (AN, γ-BL, MF, PC, DME), their mixtures, and of further additives (CCl_4, diglyme, THF, bromobenzene, pyridine and others), using DTA. DME and PC exhibited the highest exothermic initiation temperatures, 425 °C and 244 °C respectively. Selim and Bro stated [479] that "any polar solvent is intrinsically reactive toward lithium". This may possibly be undetectable by static experiments but is crucial in deposition and reanodization experiments (see Sect. 10.5.). Either lithium or aluminium can be plated from non-aqueous mixed lithium-aluminium electrolyte solutions, depending only on the composition of the solution; this amazing fact led Peled [491] to suggest that alkaline and alkaline earth metals are always protected by a film formed by reaction with the electrolyte controlling corrosion and deposition-dissolution processes.

It is now generally accepted that the success of lithium batteries is mainly due to the formation of a protective film [492]. The strategy for improving compatibility which, above all, is essential for secondary batteries is the search for stable solvents or additives which form appropriate films. It should be stressed that these investigations require highly pure solutions [492].

The positive material (anode) may induce unwanted film formation on lithium. This is observed [483] for cells with THF as the solvent and V_2O_5 or Ag_2CrO_4 as the positive materials. THF may be oxidized at V_2O_5 reacting to form a living polymer, which diffuses to the lithium electrode and reacts there to give an insoluble gel. Further problems arise from insufficient compatibility with the anode, e.g. solubilization of the positive materials by complexation [229,230,493] and from the lack of compatibility of electrolyte compound and solvent, e.g. exchange reactions of inorganic acid esters with the anion of the electrolyte [488,494].

10.4 Non-Aqueous Primary Cells

10.4.1. Commercial Cells with Solid Cathodes and Organic Solvents

This topic has been comprehensively reviewed by Scrosati[443], Dey[483], Besenhard[444], and Kronenberg and Blomgren[457]. From the numerous systems investigated the following are produced commercially for use in portable electronic devices[458, 495–498]:

$Li/LiClO_4$ (THF/DME)/CuS	SAFT (F); Dupont (USA)
$Li/LiBF_4$ (γ-BL)/$(CF_x)_n$; x ≈ 1	Matsushita (J)
$Li/LiClO_4$ (PC)/Ag_2CrO_4	SAFT (F); Mallory (USA)
$Li/LiClO_4$ (PC/DME)/MnO_2	Hitachi (J); Sanyo (J); Varta (G)
$Li/LiClO_4$ (PC/THF) or $LiAsF_6$ (MF)/V_2O_5	Mallory (USA); Honeywell (USA)
$Li/LiClO_4$ (THF or dioxolane)/CuO	SAFT (F)
$Li/LiClO_4$ (PC/DME)/Bi_2O_3	Varta (G); see Fig. 16

The cell reactions are[444,457,459,483]

$$2\,Li + 2\,CuS \rightarrow Li_2S + Cu_2S;\quad 2\,Li + Cu_2S \rightarrow Li_2S + 2\,Cu$$

$$n\,Li + (CF)_n \rightarrow n\,LiF + n\,C$$

$$2\,Li + Ag_2CrO_4 \rightarrow Li_2CrO_4 + 2\,Ag$$

$$Li + MnO_2 \rightarrow LiMnO_2$$

$$4\,Li + V_2O_5 \rightarrow V_2O_3 + 2\,Li_2O;\quad Li + V_2O_5 \rightarrow LiV_2O_5$$

$$12\,Li + Bi_2O_3 \rightarrow 2\,Li_3Bi + 3\,Li_2O\,.$$

Up to 1980 more than 350,000 Li/Ag_2CrO_4 cells and more than 100,000 Li/CuS cells have been produced[452]. Some characteristic data: open-circuit voltages 2 to 3 V; energy densities 160 to 200 W h/kg; Li/CuS has a zero self-discharge in seven years[452]; Li/MnO_2 (Varta) is available in sizes ranging from 30 to 1000 mA h[499].

For some further commercial systems[458] the battery electrolyte cannot be quoted: Li/MnO_2 (Toshiba (J); UCC (USA); SAFT (F); Ray-o-Vac (USA)); $Li/(CF_x)_n$ (Eagle Pitcher (USA); Yardney (USA)); Li/Bi_2PbO_5 and Li/CuO (SAFT (F)); Li/FeS_2 (UCC (USA)); open-circuit voltages: ~3 V (Li/MnO_2 and $Li/(CF_x)_n$); ~1.5 V (all others).

10.4.2 Commerical Cells with Liquid Cathodic Materials

The cells of this type generally use SO_2 or $SOCl_2$ as the depolarizers. The schemes of cell reaction are

$$2\,Li + 2\,SO_2 \rightarrow Li_2S_2O_4 \qquad \text{Ref. [500]}$$

$$4\,Li + 2\,SOCl_2 \rightarrow S + SO_2 + 4\,LiCl \qquad \text{Refs. [501,502]}\,.$$

These cells exhibit the highest energy and power densities down to low temperatures among the primary cells as a consequence of high conductivity, fast-reacting liquid depolarizer, large surface area of carbon electrodes[483] and low temperature coefficients of conductivity and viscosity. Hazardous currents up to 100 A were reported for these cells [483].

The Li/SO_2 battery (Mallory (USA) [443]; Duracell (USA) [443]; Power Conversion Inc. (USA) [483,503]; Honeywell (USA) [483]) uses electrolytes which typically contain 1.8 M LiBr in a mixture 23:10:3 of SO_2, AN, and PC. The system is under pressure and has a safety vent [483]. The base electrolyte solution, LiBr/AN/SO_2 shows high conductivity with a low temperature coefficient (specific conductance [483]: $\sim 5 \times 10^{-2}$ Ω^{-1} cm^{-1} (25 °C); $\sim 2.4 \times 10^{-2}$ Ω^{-1} cm^{-1} (—50 °C)); the open-circuit voltage is about 2.9 V [503] capacities range from 0.5 to 30 A h [500,503] at energy densities of about 290 W h/kg (0.4 W h/cm^3) [483,505]. Linden and McDonald [500] published a review about this battery which also summarizes the fields of possible applications.

A detailed study of the reaction mechanism [506,507] shows a remarkably high rate constant for the electron-transfer reaction, 1.0 ± 0.2 cm/s in DMF [506], and the strong influence [507] of both the supporting electrolyte and the solvent on the reduction which is explained by ion-pairing. Safety studies [486,490,508–510] show the importance of the composition of the electrolyte solution. The exothermic reaction of lithium with AN is retarded by SO_2 and PC; SO_2 did not react with lithium at 320 °C [508]. Dey and Holmes [486] published a list of eight electrolyte solutions which have higher conductivity than the standard LiBr/AN/SO_2 solution and increased stability with lithium.

The Li/$SOCl_2$ cells [511–513] (Mallory (USA), GTE (USA), Honeywell (USA) [483]) use $LiAlCl_4$/$SOCl_2$ as the electrolyte solution; the open-circuit voltage is 3.6 V; available capacities range from 0.8 to 10.8 A h [514] at energy densities of about 660 W h/kg [483,492]. Studies on the cell reaction and further aspects of the chemistry in the Li/$SOCl_2$ cell are found in Refs. [483,501,502,515–517], conductivity and viscosity studies on $LiAlCl_4$/$SOCl_2$ in Refs. [518,519].

Li/$SOCl_2$ cells may be operated at high discharge currents. Cells with discharge rates up to 300 mA/cm^2 are reported in an investigation on the cell performance [520]. The most effective discharge characteristics were obtained with 1.5 M $LiAlCl_4$ and 4.5 M $AlCl_3$ solutions [520], the highest average voltages with cells containing only $AlCl_3$ or an excess of $AlCl_3$ which dissolves the protective LiCl film of the cathode and thus limits the use of this solution to reserve cells. Reserve cells (Sonnenschein (Tadiran/GTE)) [514] work without voltage delay. Another way to circumvent voltage delays is to use lithium closoboranes, $Li_2B_{10}Cl_{10}$, $Li_2B_{12}Cl_{12}$, or other appropriate electrolytes [521].

10.4.3 Recent Developments

Some types of cell which introduce new principles should be mentioned. An organic material, pyromellitic dianhydride/pyromellitic acid, is used as the cathode material in a lithium battery [522] with 1 M $LiClO_4$/PC electrolyte solution and yields an energy density of about 1300 W h/kg. The open-circuit voltage is reported to be 3.1–3.2 V.

SO_2Cl_2 is used as a promising depolarizer in the cell Li/$LiAlCl_4$, SO_2Cl_2/C [483,523]. The suggested cell reaction is $2\,Li + SO_2Cl_2 \rightarrow 2\,LiCl + SO_2$. The cell characteristics are comparable to those of Li/$SOCl_2$-cells.

The cell Li/5 M S (as Li_2S_n), 1 M $LiAsF_6$ (THF)/C [524] contains S_n^{2-} as the depolarizer; energy density 300 W h/kg. The solubility of polysulfides in aprotic solvents is controlled by the solvent basicity [525].

10.5 Secondary Batteries

In contrast to primary cells, no secondary cell has yet reached the commercial stage. Only one secondary battery, Li/TiS_2 [526)], gives a warranty for a minimum of five charge-discharge cycles [444)]. The battery was introduced [492,527)] by Exxon. The reason for this far less successful situation is mainly due to the poor cycling behaviour of the lithium anode.

Lithium may be plated at ~100% efficiency from several non-aqueous solutions [479,492,528,529)] but is stripped anodically with much lower efficiency [492,528,530)]. Brummer et al. [492)] gave an explanation of this fact: The formation of protective films isolates lithium grains from the substrate which cannot then be discharged. This causes increasing irregularities during the following cycles by making more and more lithium electro-inaccessible [530–533)].

10.5.1 Improvements of the Cycling Efficiency of the Anode

With regard to their model Brummer et al. proposed [492,528)]:

i) the use of surface active additives as "levelling agents" or the use of "precursors"
ii) the use of alloying substrates,
iii) the use of scavengers generated internally,
iv) rigorous purification of electrolytes and solvents,
v) modification of solvent reactivity.

From these suggestions only those concerning the electrolyte solution, i) and v), will be discussed here, for further information the reader is referred to the comprehensive reviews of Brummer et al. [492,528)].

Brummer et al. [492)] quote Rhodamin B sodium salt and disodium fluorescein as examples of levelling agents. When added to $LiAlCl_4$/PC solutions the adherence of lithium is improved. Alkyl aryl polyethers or alkyl aryl polyethyleneglycol ethers serve the same purpose [534)]. Addition of tetraglyme (1–2%) to a 1 M $LiClO_4$ solution in PC containing Bu_4NI (0.6 M) results in a significantly better cycling behaviour [535)]. Macroheterocyclic compounds as additives reduce the solvent decomposition as shown by Soffer [536)] for the reaction of lithium amalgam with solutions of $LiClO_4$ in PC. Precursors [492,530,533)] are oxidizing agents which produce, in a faster reaction with the substrate than the solvent does [492)], an appropriate type of film which must be highly permeable for Li^+ but impermeable for the solvent. Brummer et al. recently reviewed [492)] their investigations on precursors in 1 M $LiClO_4$/PC [530,532)], 1 M $LiAsF_6$/MA, and 1 M $LiClO_4$/MA [531,533)] solutions. In 1 M $LiClO_4$/PC solutions, 0.1 M $PSBr_3$ and 0.01 M $POBr_3$ yielded average cycling efficiency of 85% in comparison to 40% for the precursor-free solutions [492)]. Koch and Young [537)], who used highly purified THF as the solvent, found that $LiAsF_6$ is the electrolyte which is least reactive with lithium. Bubbling N_2 or O_2 through the solution enhanced the cycling efficiency for about 10 cycles, in contrast to bubbling argon, CO_2 or to non-gased solutions. The effect of N_2 was explained by Li_3N formation which is a good Li^+-ion conductor [432)]. The AsF_6^- ion itself fulfils the conditions for being a precursor [492)] by forming brown films of $(\text{—As—O—As—})_n$ polymer and LiF [538)] in THF, but slower in the presence of N_2 and O_2. The cycling efficiency is reported to be 85.2% in agreement with the result from $POBr_3$ [492]. These examples also illustrate the necessity for highly pure solutions.

"Modification of the solvent reactivity" aims at the search for polar solvents which are less reactive with lithium. Actually, substitutions are introduced into the solvent molecule to lower its polarity. The fact that the (C—O) group (ethers) is less polar than (C=O) (PC, MA) or (S=O) (DMSO) and that cyclic molecules are more stable with lithium than open-chained ones [488,489] suggests that cyclic ethers should be the most suitable solvents for rechargeable lithium batteries [492]. A comprehensive investigation on THF and its alkylated analogues [527,539,540] yielded important information. In static tests 2-Me-THF and 2,5-di-Me-THF proved to be more stable with lithium at elevated temperatures than THF and 3-Me-THF. Solutions of 2-Me-THF containing $LiAsF_6$ were stable over a period of 13 months, THF solutions reacted after 25 days and THF solutions without the precursor, AsF_6^-, even reacted within 3 days. Conductivities of 1 M $LiAsF_6$ solutions with THF (13.7×10^{-3} Ω^{-1} cm^{-1}) and 2-Me-THF (3.0×10^{-3} Ω^{-1} cm^{-1}) as the solvents differ considerably [527] in spite of almost equal solvent viscosities (0.461 cP (THF) [541]; 0.457 cP (2-Me-THF) [542]), probably due to ion association ($K_D(LiBPh_4/THF)/K_D(LiBPh_4/$ 2-Me-THF$) = 4.4$ at 25 °C) [541,542]. The cycling efficiency of these solutions is compared in Fig. 19.

Using 1.1 C/cm^2 plating and stripping cycles on Ni, in 1 M $LiAsF_6$ solutions only 7% of the lithium is encapsulated on the 10th cycle when 2-Me-THF is the solvent, but 80% when THF [540].

Using the $LiAsF_6$/2-Me-THF solution Brummer et al. [528] have investigated three promising secondary lithium anode/intercalation cathode cells (cf. Fig. 20); the cathode materials are TiS_2, $Cr_xV_{1-x}S_2$ and V_6O_{13}. It is claimed that such cells working for 100 to 200 cycles at attractive energy densities are feasible. Additional information can be found in a recent publication [543].

Koch et al. disclosed [544,545] a DEE-based electrolyte solution, 2.5 M $LiAsF_6$ in DEE/THF (9:1), which enables bright lithium deposits up to 10 C/cm^2 by plating. This non-aqueous electrolyte solution exhibits the best cycling efficiency (>98%)

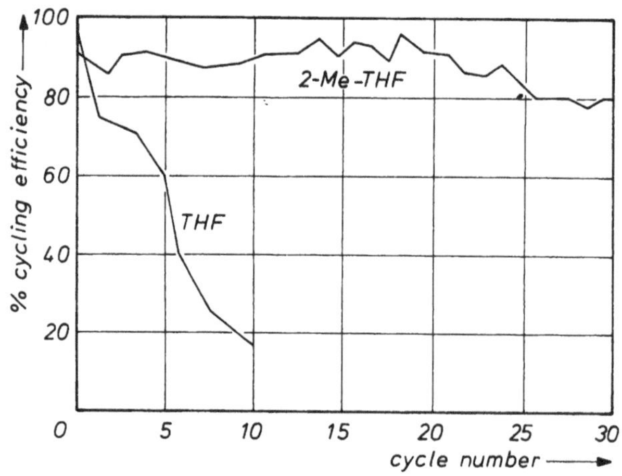

Fig. 19. Efficiency of cycling lithium on a nickel substrate from 1 M $LiAsF_6$ solutions in cyclic ethers. Anodic and cathodic current densities: 1 mA/cm^2 (with kind permission of the authors [540])

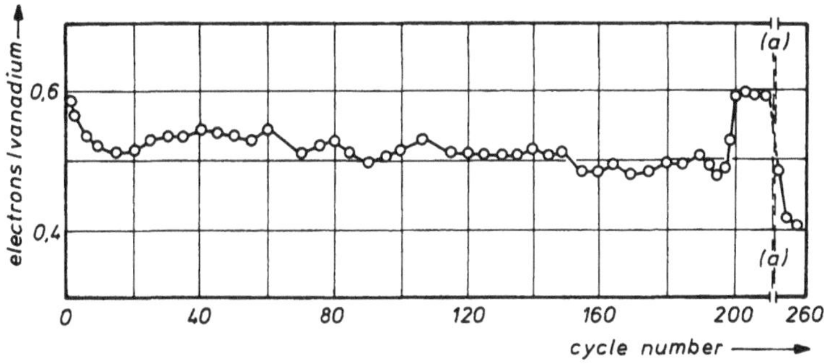

Fig. 20. Cathode utilization vs. cycle number for a hermetic laboratory test cell: Li/2-Me-THF, $LiAsF_6$/V-oxide, C. Curent density: 1 mA/cm^2; cathode capacity: 500 mA h (based on 1 electron/vanadium); current reduced to 0.5 mA/cm^2 after about 200 cycles; (**a**): interrupted scale. (with kind permission of the authors [543])

so far reported. The authors assume that the excellent stability with lithium is due to lithium ethoxide film formation.

10.5.2 Cathode Materials for Secondary Lithium Batteries

With regard to the importance for reversible lithium batteries and looking at the possible applications of reversible cathode materials in electrochromic (Sect. XI) and photo-electrochemical (Sect. XII) cells some significant results of this rapidly growing field should be presented. Both inorganic and organic materials are investigated.

Inorganic Materials with a host lattice structure can reversibly intercalate Li^+ ions, e.g.

$$xLi^+ + TiS_2 + xe^- \rightarrow Li_xTiS_2\,; \quad 0 < x \leqq 1\,.$$

The thermodynamic and kinetic aspects of intercalation processes were recently reviewed [546,547]. Materials which exhibit both good electronic and Li^+-ion conductivity, no phase change, and negligible volume effects at the intercalation process are possible candidates for new cathode materials. Scrosati [443] recently reviewed TiS_2, V_6O_{13}, $NbSe_3$, Mo_8O_{13}, and Li_xCoO_2. For further recent information including cyclabilities in non-aqueous cells with lithium anodes, diffusion coefficients, and structure determinations see: TiS_2 [526,548–550]; TiS_3 [548]; V_6O_{13} [551–553]; WO_3 [553]; MoO_3 [553]; V_2O_5 [553,554]; MoO_n with $2 \leqq n \leqq 3$ [555]; Li_xCoO_2 [556]; $NbSe_3$ [554,557]; Na_xCrS_2 with $x \geqq 0.1$ [558]; FeV_3O_8 [559]; $CuCoS_4$ [560–562]; V_2S_5 [563]; MoS_2 [564]; MoS_3 [565]; and $KFeS_2$ [566].

Some remarkable results must be mentioned in connection with non-aqueous electrolyte solutions. The use of melts of LiI·glyme solvates as the electrolyte solution in the Li/TiS_2 cell prohibits its rechargeability [550], in contrast to $LiClO_4$/dioxolane solutions: The solvation of Li^+ is supposed to be sufficiently strong to entail glyme co-intercalation. The cell of lithium/low temperature amorphous molybdenum disulfide is highly reversible with $LiClO_4$/dioxolane as the electrolyte solution: after 244 cycles the capacity still exceeds 50% of the second discharge step [564]. Dioxolane/DME mixtures are the most favourable and THF/DME mixtures the poorest solvents with regard to the performance of the cell $Li/LiClO_4$ (mixed solvent)/$CuCoS_4$ [561]. The expectation that cyclability would be improved by the use of two intercalation electrodes with different chemical potentials of the Li^+ ion as the electrodes was confirmed by Lazzari and Scrosati [567]: With a cell $Li_xWO_2/LiClO_4$ (PC)/TiS_2, open circuit voltage 2.0 V, they did not find cell deterioration upon prolonged cycling.

Organic Materials are new in this field of application. Polyacetylene, $(CH)_n$, may be chemically doped either p- or n-type [568–570] and electric conductivity can be varied over 12 orders of magnitude up to $10^3\ \Omega^{-1}\ cm^{-1}$. Golden films of composition $[CHI_{0.07}]_n$, $[CH(ClO_4)_{0.0645}]_n$, and $[CH(AsF_4)_{0.059}]_n$ with conductivities of 9.7, 970, and $260\ \Omega^{-1}\ cm^{-1}$ are obtained by the oxidation of $(CH)_n$ in KI/H_2O, Bu_4NClO_4/CH_2Cl_2, and Bu_4NAsF_6/CH_2Cl_2 solutions, respectively [571]. Polyacetylene is believed to exist as the polycarbonium ion, $(CH^{z+})_n$, stabilized by monovalent anions A^- to give $[(CH^{z+})\ A_z^-]_n$.

A non-optimized cell [571] with lithium anode and $LiClO_4$/PC electrolyte solution showed no change in its open circuit voltage (4.1 V) after 326 successive constant-current cycles. The cell with $[CH^{+0.06}(ClO_4)^-_{0.06}]_n$ cathode, energy density 176 W h/kg, is expected to give the possibility of producing inexpensive lightweight batteries with a variety of applications [571].

For further information on non-aqueous lithium batteries the reader is referred to conference reports [462, 476, 504, 572, 573], reviews [230, 438, 443–445, 457, 483, 574, 575], books [229, 576, 577], and compilations of the patent literature [434, 578, 579].

XI Non-Emissive Electro-Optic Displays

11.1 Comparison of Methods

The common features of flat non-emissive electro-optic displays actually under investigation are unrestricted viewing angle, open-circuit memory (at least transient) and hence low power consumption (30 mJ cm^{-2} [580] to 3 μJ cm^{-2} [581]), operation at low voltages and improved optical contrast, suitable threshold [580] and switching voltage levels, fast response time (< 10 ms [581, 582]; < 80 ms [580]) and long lifetime (> 10^7 cycles) [583]. The underlying controlling effects are the electrochromic effect, the electrophoretic effect, or reversible thin-layer electrodeposition.

In *Electrochromic Displays* the electrochemically induced reversible colour change, electrochromism, is accomplished either by reduction or oxidation. The extensively-studied coloration mechanism of WO_3 [584–600] can be sketched out as a simultaneous injection of cations and electrons, e.g. $ne^- + nX^+ + WO_3 \rightleftarrows X_nWO_3$; $X = H^+, Li^+, Na^+$; $n \sim 0.1–0.2$ [599]. The colour of the resulting tungsten bronze is blue, independent of the cation.

Mixed conductivity, ionic and electronic, is a feature of electrochromics. Both transport rates affect the kinetics of the electrochromic reaction [601]. The coloration kinetics and the stability of the electrode are strongly influenced both by the electrolyte solution and the preparation of the electrode material.

Other inorganic electrochromic materials studied include MoO_3 [586, 600], MoO_3/WO_3 [590], V_2O_5 [602], RhO_2 [603], Nb_2O_5 [604], and iridium oxides [580, 601, 605–608]; organic electrochromics include the rare-earth diphthalocyanine complexes [609,610] and the viologen derivatives [586,611,612] and other heterocyclic compounds.

Both solid-state [613,614] and wet cells are studied.

Electrochromic displays were claimed to be fundamentally unsuited [615] for matrix addressing. However, in his "Theory of Electrochemical Memory" Beni [616] stressed that matrix addressing of electrochromic displays should be possible by the choice of appropriate electrolytes which control both threshold and short-circuit memory. The number of lines which can be matrix addressed sequentially is shown to be in the range 10 to 10^5.

The *Electrophoretic Displays* make use of the migration of charged pigment particles, suspended in an inert solvent (generally a solvent of class 7) which contains a contrasting dye [581,617-619]. Application of a high intensity electric field drives the particles (TiO_2, anatase) carrying adsorbed ionic species (from added Na laurylsulfate, e.g. [620]) to the oppositely charged electrode. The electrolyte also acts as a dispersant.

Displays based on *Electroplating Reactions* are seldom found in the literature [582,617,621]. In these devices, reversible plating on to transparent electrodes is used.

A comparison of electro-optic displays by Pankove [622] includes systems such as light-emitting diodes (LED) and liquid crystal displays (LCD).

11.2 Displays with Non-Aqueous Electrolyte Solutions

Investigations on the XWO_3/WO_3 system in non-aqueous solutions were made with the aim of avoiding corrosion of the electrode by dissolution [591-594] as was observed in aqueous sulfuric acid solutions. Systems exhibiting Li^+ insertion from non-aqueous electrolyte solutions, e.g. $LiClO_4$ in AN [592] or PC [591] were found to be irreversible [592] or very slow [591], the response times being up to 40 s [591]. The advantages of non-aqueous electrolyte solutions are a large liquid range and the better memory [591] of the oxygen-insensitive product accessible in these solvents, namely Li_xWO_3. Only a few attempts managed to obtain the advantage of high stability provided by the non-aqueous electrolyte solutions and to avoid at the same time the drawback of slow response. The evaporated WO_3 film electrode was found to be very stable in H_2SO_4/glycerol (1:10) for more than 5×10^6 cycles at 0.5 Hz over 6 months [592]. More than 10^7 cycles and a response time of about 0.5 s were obtained with a highly porous WO_3-film and 1 M $LiClO_4$/PC solutions [623]. High Li^+-ion diffusion coefficients (10^{-7} cm^{-2} s^{-1}) in $Li_yM_xWO_3$ (M = Na, K; x ~ 0.6) were determined [624] from measurements on the cell Li/$LiAsF_6$ (non-aqueous solvent)/$Li_yM_xWO_3$ showing the influence of structure when compared to those (2.8×10^{-11} to 2.4×10^{-12} cm^2 s^{-1}, varying with x) for Li^+ in Li_xWO_3 [595].

Iridium oxide is a new electrode material with promising properties: good open-circuit memory (loss ~10% in 5 h) [607], very fast response at ambient temperatures (<40 ms) [606,607,625] and still a suitable one at low temperature (<1 s; —25 °C) [625]. This electrochromic was mainly investigated with aqueous electrolyte solutions [580,605-607,626] and solid electrolytes [627]. Non-aqueous electrolyte solutions, $LiAsF_6$/2-Me-THF and $NaAsF_6$/PC, were used for the insertion of Li^+ and Na^+ by McIntyre et al. [601]; the reported response times are 10 to 20 s.

Reichmann and Bard[604] recently reported studies on the Nb_2O_5 electrode. Using a 0.8 M $LiClO_4$/AN solution Nb_2O_5 could be coloured and bleached over three days without deterioration of the electrode at a response time of 1–2 s, in contrast to an aqueous 1 M H_2SO_4 solution where hydrogen evolution was observed. Organic electrochromics, e.g. the viologen derivatives, exhibit a remarkable stability also in organic solvents. For example, a 0.1 M solution of N,N'-diheptyl-4,4'-bipyridinium dibromide in a mixture of PC and EG (9:1) containing suspended TiO_2 is claimed to be stable for 7×10^7 cycles at 0.3 Hz at a contrast ratio of 20:1[612].

Further information on electrochromics can be found in a recent review article by Beni[583].

Electrophoretic Displays with response times of 10 ms, electrical resistivities of 10^{10} Ω cm and switching energies of 3 μJ/cm^2 at an applied field of 10^4 V/cm were obtained with TiO_2/xylene suspensions. Improvements can probably be made by an appropriate choice of the charge controlling electrolyte (ionic surfactant)[581] which stabilizes the suspension. Basic information on the conductance behaviour of these large ions in solvents with a very low dielectric constant is given by an investigation on sodium lauryl sulfate and sodium di-2-ethylhexylsulfosuccinate (degree of association $\sim 10^7$). The transient current is strongly dependent on the applied voltage conditions; the steady-state current exhibits non-ohmic behaviour, ionic dissociation and recombination processes being the controlling factors[620]. Further information on the state of technology is given in the reviews[618,619] by Dalisa. The low cost of fabrication and the high contrast over a wide range of viewing angles make electrophoretic displays promising for flat large-scale devices.

Reversible Electrodeposition of a silver iodide complex from a solution of 0.3 M AgI and KI or RbI, and I_2 in DMSO or diethyl malonate is an example for the third class of wet non-emissive electro-optic displays[582]. As long as the silver content of the solution is high enough the solvent did not deteriorate when pulses of 50 V were passed through. The addition of Al_2O_3, for preventing TiO_2 from agglomeration, and the use of $RbAg_4I_5$ in DMSO as the solid ion-conductor established a cell which survived more than 10^7 cycles when operated at <2 V drive, the realized response times were <10 ms.

XII Photo-Electrochemical Cells

12.1 Introduction

The application of the photovoltaic effect for the generation of electrical energy using sunlight as the energy source has reached the stage of commercialisation.

Investigations of non-aqueous electrolyte solutions for application in photo-electrochemical liquid junction cells have only just begun and no predictions concerning their use in commercial cells can be made at present. The better-developed[628] solid-state technologies have resulted in three lines of approach, based on different photovoltaic materials. These were described recently by Johnston[629] in his monograph "Solar Voltaic Cells" as "established technologies" for different requirements: single or semi-crystal Si cells, single crystal GaAs cells and CdS/Cu_2S cells; the attained efficiencies and fill factors ($(I_m \times V_m)/(I_{sc} \times V_{oc}) < 1$; I_m, V_m: current and voltage at point m of the I vs. V-diagram for which $I \times V$ is maximum; I_{sc}: short circuit current, V_{oc}: open circuit voltage) are: 15.5, 0.76 (Si); 20.5, 0.81 (GaAs); and 8.6, 0.71 (CdS/Cu_2S), respectively [630].

12.2 Liquid-Junction Cells

Two possible configurations are currently being investigated [631-634], the regenerative and the storage cell. The regenerative cell, where no net overall reaction occurs in the undivided liquid junction containing a redox couple, is used for the direct conversion of solar energy to electricity. The storage cell, where a membrane separates the catholyte and anolyte, converts the solar energy *via* electrochemical redox reactions and stores it as chemical energy. A new approach is energy storage by photo-intercalation [635] using layer-type compounds, e.g. p-$ZrSe_2$, in analogy to those in secondary lithium batteries (cf. Sect. 10.5.2). A variant of the storage cell can be used for producing desired chemicals by photo-electrosynthesis [628, 636].

The regenerative cell contains a narrow band-gap semiconductor, generally of n-type in aqueous cells, and a metal counter-electrode. p-Type semiconductors cannot be stabilized in aqueous solutions by redox couples [634], because the highly negative redox potentials needed would entail the reaction with water. The liquid junction is an electrolyte solution of high conductivity which contains the redox couple, e.g. S^{2-}/S_2^{2-}, Se_2^{2-}/Se^{2-}, or $[Fe(CN)_6]^{3-/4-}$. The illuminated electrode drives the majority carriers, e^- for n-type, through the semiconductor and the external lead to the counter-electrode where reduction occurs. The minority carriers, h^+ for n-type, move towards the semiconductor-electrolyte interface where oxidation at the photo-anode takes place. We are not aware of any investigations on storage cells with non-aqueous electrolyte solutions. So the following discussion is limited to regenerative cells.

The attractive feature of regenerative cells are the perfect phase contact by the liquid junction [628, 637] and the elimination of the energy loss due to absorption in the semiconductor of solid-state cells (electrolyte solutions are transparent to the major part of the solar spectrum [636, 637]). High purity of electrode materials is less important in liquid cells [632, 636]; polycrystals and thin films may be used [628, 636]; the formation of the Schottky-like barrier is technically simpler and hence cheaper than in solid-state cells [628, 632].

As efficiencies of power conversion are begining to catch up those of dry photovoltaics, several authors have claimed that economically viable liquid-junction cells which challenge solid-state cells [632, 636] will become feasible. An economic analysis based on standardized procedures was published recently [638]. It is claimed that liquid-junction cells have already at the time being the potential to meet and exceed the goal required for 1986 by the U.S. National Photovoltaic Program of a cost of 50 ¢ per peak watt, cf. Ref. [629].

Some problems have to be solved relating to the stability of the photo-anode (photo-degradation) [628, 633, 639, 640] and to the cell performance [638]. Stability criteria for the choice of appropriate redox couples [641-643] were recently discussed by

Gerischer [633]. Non-aqueous solutions provide the advantage of great flexibility for the choice of suitable redox pairs [644,645], wide anodic and cathodic stability ranges [646], and change of attackability [635,636]. The stabilization ratio (= redox current/corrosion current) varies in the order of magnitude of 10^3 and depends strongly on the organic solvent for the stabilization of n-CdTe with ferrocene [647]. For further aspects see Refs. [635,648–657], especially for the use of the more stable transition metal dichalcogenides which exhibit d-d-phototransitions without splitting chemical bonds of the semiconductor.

12.3 Non-Aqueous Solution in Liquid-Junction Cells

In aqueous cells the best performances of n-type CdS, CdSe, CdTe and GaAs photo-anodes were obtained by stabilization with chalcogenide/polychalcogenide redox couples. Noufi and Tench [658] reported an efficiency of 12%–14% using n/n^+ GaAs in selenide/polyselenide solutions for polycrystalline and single crystals of GaAs; the fill factors were 0.67 and 0.8 respectively. The photocurrents in these systems, however, deteriorate with time, especially at high light intensities [659]. Strong specific adsorption of the chalcogenide and the slow two-electron transfer are considered to be reasons for the limited stability [659].

The main advantages of non-aqueous cells are the flexibility for choosing suitable [644] and less noxious [645] redox pairs, changed attackability, and wider stability ranges (e.g. AN: 5 V; H_2O: 1.5 V) [646] as already mentioned. Many one-electron redox couples can be photo-oxidized on n-GaAs or photoreduced on p-GaAs at less negative potentials than on Pt [644]. The search for the new and promising intercalation electrodes for storage cells [635] is facilitated by the multitude of possible solvents.

A drawback is the lower conductance of non-aqueous solutions entailing lower fill factors. The investigation of non-aqueous electrolyte solutions for applications in photoelectrochemical cells is still in its swaddling cloths. Only few aspects can be discussed.

The resistance of n-CdSe to photodecomposition varies with solvent in the following order: AN $\approx$ PC $>$ DMF $>$ EtOH $\approx$ MeOH $\gg$ H_2O [659]. Methanol/$[Fe(CN)_6]^{3-/4-}$ was chosen for subsequent stability tests because of the high rates of electron transfer and its transparence to most of the solar spectrum. The conversion efficiency for the photo-anode at 85 mW/cm^2 (tungsten-halogen illumination) is reported to be 3.5% at a constant current density of 6 mA/cm^2 and the fill factor is 0.56 [659]. The anode was completely stabilized by the redox couple and no deterioration of the surface or degradation of the photoresponse were observed [659]. Specific adsorption at the electrode surface and ion pairs in the bulk solution limit the maximum short-circuit current to 17.5 mA/cm^2 in stirred and 7.5 mA/cm^2 in unstirred solutions [660]. Purity of the solvent is essential [659].

Langmuir et al. studied the stabilization of n-GaAs photo-anodes by some redox couples (Br_2/Br^-, I_2/I_3^-, I_3^-/I^-, and ferrocene, acetyl ferrocene, N,N′-tetramethyl-p-phenylenediamine with their appropriate oxidated species) in 0.01 M $LiAsF_6$ solutions of PC [645]. With the exception of Br_2/Br^-, which corroded GaAs even in the dark, these redox couples proved to be suitable. The low efficiencies (up to 7%)

obtained are related to low redox-pair solubilities and low conductivities of the solution. Anthraquinone, p-benzoquinone, ferrocene, dimethylferrocene, hydroxymethylferrocene, and N,N'-tetramethyl-p-phenylenediamine were used as depolarizers in n-GaAs/0.2 M Bu_4NClO_4 (AN)/Pt liquid-junction cells [644]. The n-GaAs was operated at potentials (0.1 to —0.6 V vs. SCE) where photodissolution would occur in water: $GaAs + 6h^+ \rightarrow Ga^{3+} + As^{3+}$ [644]. Using the ferrocene/ferrocinium couple and 0.52 mW/cm^2 of 720–800 nm radiation a maximum energy conversion of 14% (with solar radiation at 9 mA/cm and 0.2 V only 2.4%) was obtained [644]. The complete removal of water is necessary to yield sufficiently high electrode stability. For information about decreased photo-dissolution of photoconductors in AN see the papers of Bard et al. on n- and p-Si [661], n- and p-GaAs [662]. n-CdS, n-GaP, n-ZnO [663], n- and p-InP [646].

The photo-electrochemical generation [664–665] of a conducting film on the photoanode is a new way to prevent photo-anodes from photodegradation. For example, an insoluble and conducting polypyrrole film (10–100 Ω^{-1} cm^{-1}) was produced on the n-GaAs anode from a 0.1 M Et_4NBF_4 acetonitrile solution containing 0.1 M pyrrole at 0.45 V vs. SCE with the help of a tungsten-halogen lamp operating at 50 mW/cm^2 [664]. The unprotected anode deteriorated in less than 1 min whereas the protected one was operated for 100 h without deterioration, see Fig. 21 [665]. Recently this method was also applied to n-CdTe, n-CdSe, n-CdS and n-Si electrodes which were studied in aqueous and non-aqueous systems [665].

For further aspects and information the reader is referred to monographs and reviews, Refs. [628,631,657,666–671].

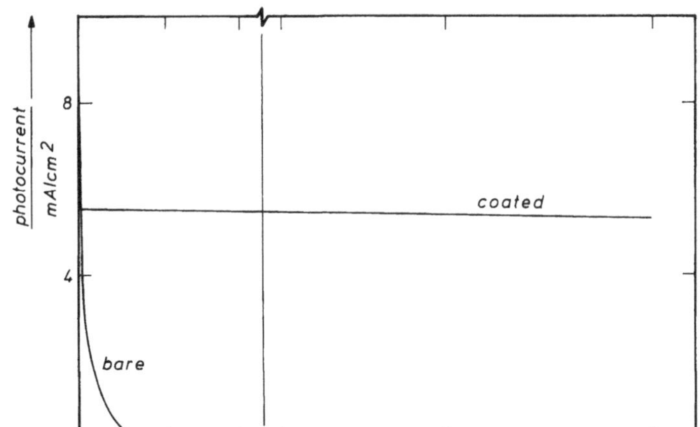

Fig. 21. Short-circuit photocurrent vs. time for bare and polypyrrol-coated n-GaAs electrodes in methanol/0.2 M $Fe(CN)_6^{3-/4-}$, 0.1 M Et_4NBF_4 solution [665].
(reprinted by permission of the publisher, The Electrochemical Society, Inc.)

XIII Electrodeposition

13.1 Introduction

At first sight, non-aqueous electrolyte solutions are not suitable for the electrodeposition of metals and compounds for many reasons:

— the high costs of solvent and solute combined with high purity requirement;
— complex operating conditions caused by toxicity and flammability;
— poor conductivity (cf. table X);
— low current efficiencies as a frequent consequence of unwanted side reactions such as the decomposition or polymerisation of the solvent which also causes impure metal deposits.

Nevertheless, numerous publications with promising contributions to this field of application show the increasing interest in non-aqueous solutions. Many metals of technological importance cannot be electrodeposited from aqueous solutions because the potentials are so negative that hydrogen evolution would occur instead of metal deposition [672-674]: alkali and earth alkali metals, B, Al, (Ga), Si, the lanthanides, and some of the transition metals (Ti, Zr, Hf, V, Nb, and Ta which are referred to as valve or refractory metals). Some metals, for which deposition from aqueous solutions is thermodynamically favourable, cannot be obtained as pure deposits: W [577,672,673], Mo [577,672,673], and Ge [577,672].

Attempts to electroplate the valve metals have so far not been very successful. Takei [674] classifies them as metals which cannot be electrodeposited from either aqueous or non-aqueous solutions because their inner d-orbitals are not filled. They are believed to form strong covalent bonds with ligands. As a consequence metal deposits contain carbon, oxygen or halides.

One commercial application is the deposition of Li from non-aqueous solutions for the use in high energy-density batteries (cf. Sect. X). Coating, especially with Al, was proposed for uranium fuel elements [675,676], steel [673,677], production of wave guides [672], high surface-area deposits for electrolytic capacitors [678], and mirrors [679]. Protective coatings with aluminium are considered to be more efficient than zinc coatings and even cheaper [673]. Electrorefining of Al in non-aqueous solutions may use up to 85% less energy than the fused-salt electrolysis [672,680] commonly used. The electrodeposition of Be has appealing aspects for the atomic energy industry [681]. A new and intensively researched field is the electrodeposition of semiconducting materials for the production of inexpensive large terrestrial solar cells [682]. This application includes the electrodeposition as thin films of both pure semiconducting metals, such as Ge or Si, and compounds, e.g. CdS and CdSe. A further field of research is the codeposition of metals as alloys by making use of the potential shifts relative to water.

Even metals being depositable from aqueous solutions may be advantageously plated from non-aqueous solutions when special surface properties are wanted, e.g. greater coat thickness or higher cathodic efficiencies [683].

13.2 Selected Examples

The electrodeposition which is commercially most important [577,673] is that of Al by the NBS process [684,685]. The plating bath is a mixed electrolyte solution of $LiAlH_4$ (0.2–0.4 M) and $AlCl_3$ (3 M) in anhydrous DEE. Attempts have been made to reduce the volatility of solvent (flammability), to improve the operating life of the bath, to increase the current density, and to find substitutes for $LiAlH_4$.

The Etmac bath [676,686] consisting of a solution of $AlCl_3$, $AlHCl_2$, and 2-ethoxyethyl trimethylammonium chloride in DEE shows reduced volatility and, furthermore, avoids the restriction through the initial aluminium content of the bath. Aluminium films up to 130 µm thick were obtained with current densities of 0.5 to 15.5 A/dm^2.

Commercial baths using THF solutions ($AlCl_3:LiAlH_4 = 5:1$) [677] are now available [681] and smooth and coherent deposits were obtained with high current densities (up to 18 A/dm^2 without stirring). Additions of benzene were investigated. A THF/benzene bath ($AlCl_3:LiAlH_4 = 3$ to 1) [687] with a total aluminium concentration $\geqq 0.7$ M shows advantages with regard to volatility, stability, and operating life when compared to the NBS-bath. Deposits were obtained at current densities of up to 10 A/dm^2 with stirring and the bath is applicable to aluminium refining. In the range of 0 to 50 Vol% of benzene, the specific conductance of these baths is almost independent of solvent composition at high aluminium contents (equimolar $LiAlH_4/AlCl_3$ mixtures, both concentrations at 0.7 M; specific conductance 4×10^{-3} Ω^{-1} cm^{-1}). For further information on solvent mixtures of THF see [688], for other solute mixtures in THF yielding conductivities up to 10^{-2} Ω^{-1} cm^{-1} see [439,689,690].

Investigations on Al_2Br_6 as the source for aluminium exhibit some interesting features for the system KBr (or LiBr)/Al_2Br_6 toluene [235,237,691–696]. The analysis of transference numbers and e.m.f. data exhibits triple ions, $(K_2[Al_2Br_7])^+$ and $(K[Al_2Br_7]_2)^-$ in equilibrium with KBr and Al_2Br_6 as the predominant ionic species [237]; the observed high conductance of 1 M Al_2Br_6/0.8 KBr (6×10^{-3} $\Omega^{-1} \times$ $\times cm^{-1}$) in an inert solvent (toluene), suggests a non-Stokesian (hopping, cf. Sect. 6.3) conductance mechanism [235].

Another way of replacing the expensive hydrides of the Etmac bath was reported by Yoshio et al. [697]. Using DEE solutions of Al_2Br_6 and LiCl with conductivities of 6 to 10×10^{-3} Ω^{-1} cm^{-1}, current densities of 20 to 100 mA/cm^2 were obtained at efficiencies of essentially 100% giving bright and adherent aluminium coatings. Takei [681] reports a bath containing $Al(CF_3COO)_3$, 400 g/dm^{-3}, in MeOH for 31% cathode efficiency. Thin and uniform Be films are obtained from $Be(CF_3COO)_2$/CF_3COOH/Na_2SO_4/AN baths [681].

Phosphotungstic acid in THF, DMF and FA with H_3BO_3, NH_4Cl, H_2O_2, and citric acid as additives were investigated for tungsten deposition [681]; X-ray analysis showed tungsten and tungsten oxides as the components of the deposits from FA solutions. From phosphomolybdic acid in FA an analogous result, Mo and Mo-oxides, was obtained [681].

Titanium is generally produced by the Kroll or Hunter process and recently also by energy saving deposition from high-temperature salt baths as known for coating [698,699]. Many attempts have been made [672,700,701] to replace the high-temperature process with a low-temperature electroplating process. Brenner et al. [701] listed more than 50 solvents, but did not succeed in plating Ti as a pure metal. Only Ti/Al deposits containing up to 6% Ti were obtained from a bath containing hydrides and borohydrides. Biallozor and Lisowska [700] obtained layers about 15 μm thick with a titanium content of 37% by electroreduction of $TiCl_4$ in acetonitrile solutions containing $LiBF_4$. The carbon content of the deposit is 60%. Santons and Dyment [702] electrolysed $TiCl_3$ in HCl/DMSO solutions and obtained adhesive and homogeneous films up to 75 μm thickness. They claimed the method to be also successful for some other valve metals (Zr, Hf, Nb).

Tantalum coatings are of great technical importance as a consequence of their good high-temperature properties ($T_F \approx 3000$ °C), excellent corrosion resistance, and high thermal conductivity. Matching surfaces with Ta sheets is used technically despite the high cost, poor heat conduction and resulting material stress [703]. Electrodeposition from salt melts wastes energy and imposes technical problems [703]. Fischer and Schwabe [703] investigated deposition from $TaCl_5$/PC solutions as a function of conductivity, temperature, electrolyte concentration, and current density. They obtained compact Ta coatings containing traces of Cl.

Electrodeposition of silicon can be achieved from PC baths containing tetra-alkylammonium chlorides and $SiHCl_3$ as the Si source [704]. Deposits on a variety of materials including low-cost substrates such as the Ti-6 Al-4 V alloy or coated fused silica were made. Both surface morphology and current decay resulting from the increase of electrical resistance of the growing Si film can be controlled by the cation size of the supporting electrolyte, R_4NCl. Bound hydrogen (SiH_2 or SiH) can be driven off at 470 °C. The amorphous Si film exhibits photoconduction and photovoltaic properties and offers an inexpensive route for solar cell applications.

Recently, the electrodeposition of amorphous silicon was achieved using EG and FA/EG solutions of HF and tetraethyl orthosilicate or silicic acid as the silicon sources [705]. The conductivity changes from p- to n-type, when doping with phosphorous. This is possible by the addition of triethyl phosphate to an EG solution of HF containing 10^{-2} M tetraethyl orthosilicate [705].

For details of the deposition of other semiconducting materials, e.g. CdS, CdSe, Ge, from molten salts and organic solvents the reader is referred to a recent comprehensive review by Elwell [682].

XIV Wet Electrolytic Capacitors

Electrolytic capacitors contain Al, Ta, Nb, Ti, or Zr foil electrodes which are electrolytically oxidized in a dielectric formation process [617, 706–710]. The oxide films are ≤ 100 nm thick [617]. The most important electrode metals are Al, Ta and Nb with the electrolytically-formed oxides [706, 711–713] Al_2O_3, Ta_2O_5 and Nb_2O_5 of static permittivities up to 10 [617], 28 [617, 714] and 41 [714], respectively. Aqueous as well as non-aqueous electrolyte solutions are used to establish the contact between the oxide film and the counter-electrode (cathode), especially in Al_2O_3 capacitors [617]. Ta_2O_5 capacitors are almost always constructed from solid electrolytes [617].

The characteristics of electrolytic capacitors are their high specific capacitance (up to 400 μF/cm²) [711] and the dielectric oxide layers withstanding the intense electric field (up to 10^7 V/cm) [617, 706]. The main drawback is the high dissipation factor, especially at high frequencies and low temperatures. The electrolyte is the controlling factor which determines dissipation [706] and shelf life [715]. Wet capacitors with non-aqueous electrolyte solutions exhibit wider temperature ranges and higher corrosion resistances than those with aqueous solutions.

The general requirements for the electrolyte solution are almost the same as those for HEBs, *i.e.* low resistivity ($300 < \varrho/(\Omega\,\text{cm}) < 2000$) [707] over a large temperature range, low vapour pressure of the solvent, and sufficient solubility of the electrolyte. Specific compatibility conditions must be fulfilled. The electrolyte should act as a good oxide-forming agent, maintaining the electrochemical state of repair of the oxide layer during the life time of the capacitor [617,708]. Glycol and glycerol with appropriate additives for reducing the water content and for stabilizing the oxide layer, and borates as the solutes are well-known capacitor electrolytes [708,715,716]. Their disadvantages are very high viscosity and temperature coefficient of viscosity which reduce the usable temperature range. Furthermore, the oxide layer is not resistant to this electrolyte at higher temperatures [715].

Improvements in capacitor electrolytes are obtained by optimizing binary and ternary solvent mixtures and using appropriate additives. Some examples may illustrate this field of application. Capacitors using N-methylformamide and N-ethylformamide solutions of triethylammonium maleate and monomethylammonium maleate were studied over a large temperature range ($-65\ °C < \theta < 85\ °C$) [717]. Solvent mixtures of PC (35.5–43.5 weight%), γ-BL (35.5–43%), and NMA (1.4 to 16.2%) with tetramethylguanidine, picric acid, and citric acid as the electrolytes [718] show increased lifetime. Low specific resistance (140 Ω cm; 50 °C) was found for glycol/DMF solutions of p-nitrobenzoic acid, neutralized by diethylamine in the temperature range $-40\ °C \leq \theta \leq 100\ °C$ [719]. Capacitors using DMF or one of its homologues and the azeotropic mixture of Et_3N and acetic acid show a long shelf-life and improved low-temperature performance [720]. P_2O_5 [715,720] and H_3PO_4 [721] were proposed as deterioriation inhibitors. For further information concerning the use of non-aqueous solutions see [617,709,710,712,716,722–724]. Information on the electric breakdown in wet electrolytic capacitors can be found in Refs. [725–730].

A new promising type of electrolyte capacitor making use of the double-layer capacitance at large surface electrodes was recently proposed [731,732]. Only capacitors with aqueous electrolyte solutions have so far been investigated. They can attain capacities up to 1 F/cm² and energy densities up to 25 W h/l. They may be used to protect memories during brief circuit interruptions instead of batteries. Because of water decomposition only voltages up to 0.9 V can be applied. Consideration should be given to the use of non-aqueous electrolyte solutions which would permit the application of higher voltages, cf. Ref. [731].

XV Electro-Organic Synthesis

15.1 Disappointments and Advantages

Swann's "Bibliography of Electro-Organic Synthesis 1801–1975" contains about 12,000 citations involving electro-synthesis reactions [733–735]. Only a handful have reached large-scale commercialization [733,736]. Based on Swann's work, Beck, Alkire and Weinberg compared the energy requirements of further possible candidates for electrochemical synthesis on the large tonnage scale (> 10^4 t/year) with the conventional routes [737].

Despite this disappointing situation, numerous examples in open and patent literature indicate continuous research activity in both the academic and industrial field. "Fossil fuels for energy and feedstocks are dwindling rapidly, environmental legislation is becoming increasingly demanding" [738], and "industrial countries are moving increasingly towards an electrical economy" [739]. The reconsideration of the possibilities made available by electro-organic synthesis is useful when taking into account the actual situation. "It seems likely, indeed, that the one-generation future will bring an electro-chemical industry which will be more than half of chemical industry and indeed it seems likely to guess that in a two-generation future chemical industry will become largely electrochemical"; this forecast [740] is very optimistic but should stimulate us to do everything possible in this field of application.

The advantages of electro-organic synthetic methods are obvious:
- pollution-free oxidation and reduction carried out without oxidizing or reducing agents and without catalysts;
- controlled reaction rates;
- high selectivity when the potential along the electrode surface is uniform [741];
- fewer reaction steps;
- simplified product recovery and purification.

Reaction intermediates such as radicals, carbonium ions, carbanions, cation and anion radicals, dications and dianions can be produced. For example, six reversible redox reactions of 9,9'-bianthryl-10,10-dicarbonitrile in 0.1 M Bu_4NPF_6/propionitrile solution yield a spectrum of products ranging from the tetra-anion to the dication [742].

The energy to overcome the reaction barrier is introduced by electrical energy, which can be finely tuned, instead of thermal energy and thus can be applied at low temperatures to thermally-sensitive compounds [739]. The processes offer great ease in monitoring the reactions and are suitable for continuous and automatic operation [743]. Spontaneous electrochemical reactions which produce chemicals and electric energy simultaneously, e.g. electrogenerative halogenation [740, 744], and electro-initiated polymerizations [746] are further promising fields.

However, parameters concerning the electrode potential, electrode material, current density, electric field, conductance, and absorption must be controlled and optimized in addition to the usual ones, *i.e.* concentrations, temperature, pressure, and time [738, 747, 748]. Scaling up of electrode reactions, which depend on the reactor surface instead of reactor volume as usual, is difficult and expensive [736, 738]. Fabrication of appropriate electrodes and cell design require new technologies [738]. Some typical examples are given for illustration.

Jansson and Tomov [749] studied the dehydrodimerisation of diethyl malonate in KI/AN using six types of cell. The cell performance depends on the proper mixing for high (>40%) conversion and on the choice of the electrolyte and solute concentrations at low (<40%) conversion. The reduction of nitrobenzene yields nine products depending only on the electrolyte solution and the electrode material [738]. Alkire and Gould [741] published an engineering model for continuous flow-through porous electrodes. They studied the multiple-reaction sequence (EEC) in the oxidation of 9,10-diphenylanthracene using 0.2 Et_4NClO_4/AN as the electrolyte solution. The excellent agreement of 4% deviation of experience and model calculation shows the benefit of such models for estimating the influence of the reaction cell parameters and for scaling up the process. Further optimization calculations for electrochemical processes and cell design can be found in Refs. [733, 747, 748, 750–753].

Recently numerous investigations have been carried out in non-aqueous solutions, many of them with mechanistical orientation, see [733, 736, 743, 745, 754–771]. The work devoted to the study of fuel cells and high-energy batteries stimulated [743] electro-organic synthesis. Appropriate solvent properties for organic reactants and reaction products, the wide range of polarity which can be achieved, the large range of electrolytic inactivity (up to $> \pm 3$ V) far beyond that of water, fast electron transfer, and new and unusual types of reaction favour the use of non-aqueous solvents. For electro-activity ranges depending on the supporting electrolyte see Refs. [41, 759, 772]; e.g., the use of supporting electrolytes with BF_4^- and PF_6^- anions in AN even permits the electrochemical study of allylic or tertiary C—H bonds [773]. Non-aqueous solvents can be selected for stabilizing reactive intermediates, for acting as a reactant or favouring a variety of desired reaction paths.

The main problems in industrial processes are the costs of highly purified solvents and the ohmic drop across the cell resulting in energy wastage. For example, the inactive range of a 0.1 M DMSO solution of $NaClO_4$ (Pt-electrode) is diminished markedly by the residual water content [774]. The drawback due to poor conductivity could be overcome by the use of mixed solvents and weakly-associating electrolytes, cf. battery electrolytes (Sect. X). Comprehensive information can be found in monographs and reviews, Refs. [733, 735, 736, 743, 745, 754–758, 760–764, 766–771, 775].

15.2 The Influence of Electrolyte Solutions on Reactions

The general conditions for the choice of appropriate solvents and solvent mixtures concerning solvent classes, permittivity, viscosity, etc. are given in Section X. Supplementary conditions may result from requirements of solution structure near to the electrode or from the properties of intermediate reaction products as the new solutes in the solution. Solvation and ion-pair formation of these species depend strongly on the electrolyte solution, cf. Section VII, and control both reaction path and reaction rate. Some examples may illustrate these features.

The reaction rate of the electrohydrodimerization of 1,2-diactivated olefines in DMF solutions is significantly increased by adding alkali-metal salts [776] to a solution containing Bu_4NI as a consequence of ion-pair formation with the intermediate radical anion. Formation of ion pairs and possibly of higher aggregates with the benzil radical anion is discussed in the electrochemical reduction of benzil in DMSO, DMF, and AN solutions containing Bu_4NClO_4 and alkali metal perchlorates [777]; voltametric measurements show fast reversible association reactions of the benzil radical anion with the alkali metal ions whereas Bu_4N^+ exhibits only weak ion-pair formation.

The cation plays an important role [778] in the electrochemical formation of the commercially important mono- and disubstituted N-alkyl formamides from CO in

MeOH and alkylamine solutions. The use of 0.2 M Et_4NBr, Et_4NBF_4, Bu_4NBr or $NaBF_4$ results in high yields (1 mol/F), that of ammonium salts gives no yield at all. Alcoholate ions, which can be formed only in the presence of R_4N^+ and Na^+, catalyze the reaction; the catalyst is believed to be highly concentrated in the diffusional electrical double layer, especially when R_4N^+ are the cations.

The reductive cyclotetramerization of CO to squarate ions has been extensively studied in DMF, HMPT, diglyme, THF, NH_3, AN and DME using Bu_4NBr as the supporting electrolyte, in DMF with $CaCl_2$, $MgCl_2$, KBr, LiCl, NaN_3 and some R_4N^+ salts and in DME with $LiClO_4$. It is another example where ionic effects occur [779]: large cations and anions give high yields, small cations and anions poor ones.

15.3 Selected Examples from Actual Investigations

Just as in the preceding sections, only a few examples from the recent literature will be given to illustrate the various types of research and application.

The electrochemical synthesis of tetraalkyl lead by the Nalco process [733,736,760,780,781] is carried out in mixed ethers as a large-scale industrial process.

The potential shortage of hydrocarbons as raw materials may favour [755,782] reactions with small molecules. Carbon dioxide is reduced to oxalic acid in aprotic solvents (DMF, AN, PC) containing tetraalkylammonium salts [783-786]. Oxalate and succinate are obtained by electroreduction of CO_2/C_2H_4 mixtures in DMF or HMPT at elevated pressure [787]. Carbon monoxide is reduced to squaric acid at high pressures in DMF, HMPT, and THF solutions of Bu_4NBr [788]. Organic carbonates at high pressure (100 atm) were synthesized by electrolysis of a solution of CO in the appropriate alcohol in the presence of various halides which also act as catalysts (e.g. NH_4Br) [789]; these important solvents are thus available without the use of phosgene and the problem of the disposal of hydrochloric acid. For further use of CO as a starting material see [778,779]. Cyclic sulfones of high purity are produced by reduction of SO_2 in AN solutions in presence of some dibromides [790], symmetric $R{-}SO_2{-}R$ [791] or non-symmetric $R{-}SO{-}R'$ [792] sulfones in presence of alkyl halides or a mixtures of two alkyl halides, respectively.

Monomethoxylated amides, used for the preparation of vinylamides, were obtained in yields up to 100% by anodic oxidation of N,N-dialkylamides in methanol solutions of Bu_4NBF_4 [793]. N—N coupling can be achieved by anodic oxidation, e.g., amide anions of secondary amines in $LiClO_4$/THF yield tetrasubstituted hydrazines [794]; monosubstituted sulfamide anions in MeOH are oxidized to azo compounds which can be cathodically reduced to symmetrically-substituted hydrazines, and imidodisulfonate trisanions to hydrazine tetrasulfonate which can be converted to hydrazine.

Heterocyclic compounds (e.g. indazoles) have been synthesized [795] by anodic oxidation of hydrazones, in $LiClO_4$/AN. Tetrazolium salts are obtained by oxidation of formazans in Et_4NClO_4/AN, see [796].

Anodic coupling of vinyl ethers, phenols and phenol ethers is claimed to be promising [797] for the synthesis of natural products and medicaments, e.g. cyclization of methoxybibenzyls in $LiClO_4$/AN [798,799]. Dimethoxydihydrofuran, a sales product for disinfectants, can be produced without pollution by methoxylation of furan in NH_4Br/MeOH. Nohe [800] has made an economic comparison with the competing conventional route and the effect of cell design on this process is reported by Jansson and Fleischmann [801]. Lelandais [754] refers to acetimidation reactions in $LiClO_4$/AN and Et_4NBF_4/AN [802, 803] as a promising route to pharmaceutically active products. The trifluoroacetoxylation of chlorobenzene in CF_3COONa/CF_3COOH is carried out [804] with 94% product yield at 70% current efficiency. The products may be hydrolyzed to the corresponding phenol, resorcinol or pyrocatechol derivatives and hence can be used as a simple way to the technically-interesting 4-chlororesorcinol. Production of dimethylsebacate by the Kolbe synthesis in aqueous methanol is a further process which has reached the pilot stage [733, 736, 800]. Studying the Kolbe-synthesis with CF_3COONa, $CF_2ClCOONa$ and $CF_2BrCOONa$ in Bu_4NPF_6/AN solutions Waefler and Tissot [805] obtained the difluorocarbene as a by-product at room temperature from $CF_2BrCOONa$, the Kolbe synthesis being the main reaction.

A pilot process for production of dihydrophthalic acid used dioxane-water mixtures [736, 738, 800, 806]. The influence of further organic co-solvents on yields is discussed by Nohe [807]. Pletcher and Razaq have accomplished [808] the electrochemical reduction of 2,3,5-tribromothiophene to 3-bromothiophene, in NaBr solution, 0.2 M in aqueous dioxane (70% dioxane), with high current efficiencies. This process could replace the conventional reduction with a three-fold excess of Zn, which leads to the disposal of waste containing Zn^{2+}. Acetophenone may be electrochemically reduced to ethylbenzene in acidified ($HClO_4$/H_2SO_4) ethanol. This is a further example for avoiding the problems allied with the use of metal powders in reductions [809].

Aqueous Bu_4NBr/trioxane solution proved to be an efficient medium for the reduction of benzene to cyclohexadiene [810]. The advantage of trioxane when compared to dioxane results from the higher conductivity of its electrolyte solutions and the possibility of applying more negative potentials.

The use of solid polymer-electrodes is a new technique in electro-organic synthesis avoiding the supporting electrolyte and the side reactions caused by it. For example, Ogumi et al. [811] hydrogenated olefine double bonds in EtOH, DEE, and hexane without supporting electrolyte.

The application of polymer-coated electrodes was already discussed for high energy batteries (doped polyacetylenes, Sect. X) and liquid junction solar cells (conducting polypyrrole films, Sect. XII). Further information on the application of polymer-modified electrodes can be found in recent reviews [812, 821–823].

Baizer has critically reviewed [767] the prospects for further application of organic electrosynthesis and has compiled a list of conditions for the successful use of this technique. The main suggestions were: use of electrochemical methods in oxidation or reduction processes, where stoichiometric concentrations of oxidants or reductants, especially exotic ones, cf. Refs. [733, 770, 806], are needed in conventional synthesis; the use of an electro-organic synthesis where it is the only successful approach; finally in small-scale processes, e.g. the production of medicinals or fine chemicals. These and further suggestions may be found in Refs. [733, 754, 769, 770]. For further information, concerning other examples of electro-synthesis, the reader is referred to recent reviews published by Lelandais [754], Fioshin [755], Baizer [767], Köster and Wendt [806], Pletcher [760], and Schäfer [770], and the volume 75 of the American Chemical Engineers Symposium Series [733].

XVI Further Promising Fields of Application

Some applications of interest which cannot be subsumed under the headings of the previous sections are still to be mentioned.

16.1 Processes Based on Solvating Properties

Parker [813] developed the energy and capital-cost-saving "roast-leach-disproportionation process" for sulfide concentrates of Cu and Ag. The process makes use of the different solvating powers of AN and H_2O, respectively. The equilibria

$$Cu + CuSO_4 \rightleftarrows Cu_2SO_4$$

$$2\,Ag + 2\,CuSO_4 \rightleftarrows Ag_2SO_4 + Cu_2SO_4$$

are shifted to Cu^+ and Ag^+ which are strongly solvated by AN when adding an acidified AN/H_2O solution. In contrast, pure water favours the Cu^{2+} ion and the pure metals. As a consequence materials containing impure copper or silver may be leached using acidified $CuSO_4/H_2O$/AN solutions; subsequent distillation of AN from the solution yields the purified metal; the waste heat of the roast process can be advantageously used for carrying out the distillation. The energy saving (>50%) and the reduction in capital cost (>60%) of this process as compared to an electrolytic copper refinery are impressive.

Using the same principles, waste materials such as cement silver, silver halide residues and circuit boards can be recovered [813] using $CuCl_2$ as the oxidant in NaCl/DMSO solutions. The anions $CuCl_2^-$, $AgCl_2^-$ and $AuCl_2^-$ are strongly solvated in DMSO in contrast to water.

$$Ag + CuCl_2 + 2\,NaCl \rightleftarrows CuCl_2^- + AgCl_2^- + 2\,Na^+$$

$$Au + 3\,CuCl_2 + 4\,NaCl \rightleftarrows 3\,CuCl_2^- + AuCl_4^- + 4\,Na^+ .$$

Addition of water to these solutions yields gold metal or silver chloride precipitates, because the Cl^- ion is strongly solvated in water. It should be mentioned that these processes do not involve pyrometallurgical steps.

The two-phase battery $Zn/ZnBr_2$, $(H_2O)/Zn(Br_3)_2$, (dipolar aprotic solvent)/insoluble electrode is a further example [813] using the stronger solvation of halide ions by dipolar aprotic solvents when compared to water. As a further advantage of different solvent solvation powers the Br_2 is kept away from the Zn electrode. The battery with an overall efficiency of about 60% at 20 mA/cm charge-discharge current densities and an open-circuit voltage of 1.76 V is claimed to have some promise for load-levelling applications.

For further aspects of solvating properties which have been used technically see Ref. [813] and literature quoted there.

16.2 Electropolishing

Electropolishing is a technique where non-aqueous solutions are applied to even out surface irregularities. A bath for polishing Mo, V, Al, Fe, and U/V alloys consists of a 1 M H_2SO_4/MeOH solution [814]; $HClO_4$ (5–20%)/MeOH can be used for Al, Cu, Pb, Sn, and Fe [815].

16.3 Anodic Oxidation of Semiconductors

Anodic oxidation is used commercially in the production of metal-oxide semiconductor (MOS) devices on GaAs [816,817]. The production of As_2O_3/Ga_2O_3 films is best carried out with glycol/water or propane-1,2-diol as the solvents. The resulting stable oxide films exhibit 10^5 to 10^8 times higher resistivities than those from aqueous anodization. Glycol can be replaced by other protic solvents, e.g. N-methylacetamide [818].

Electrochemical etching and electropolishing of semiconductors are further techniques which should be mentioned. The reader is referred to Ref. [819].

XVII Acknowledgements

We are very grateful to the Herbert Quandt Stiftung who sponsored our investigations by a grant and to the DECHEMA, Projekt: Entwicklung computergestützter Rechnungsmethoden für Stoffdaten, which is financially supported by the Bundesministerium für Forschung und Technologie.

XVIII Appendices

A Solvent and Electrolyte Data

A.1 Properties of Organic Solvents

Reliable data of organic solvents are needed for the control of purity as well as for the parameters in the equations expressing the concentration dependence of the properties of electrolyte solutions. Table A-I contains a selection of currently used solvents which are arranged in classes according to the principles given in Section II, cf. also Table I. Data are given for 25 °C if not indicated otherwise.

Improvements of the purification processes and new methods of purity control have provided data which often deviate significantly from the traditionally used ones. As far as we are aware of trustworthy new data these are given in Table A-I.

Table A-I. Properties of organic solvents

Solvent	Abbreviation	θ_F / °C	θ_B / °C	ε	η / cP	ϱ / g ml^{-1}	DN [110,130] / kcal mol^{-1}	E_T [111] / kcal mol^{-1}	AN [112–114,130]
1. Amphiprotic [103] hydroxylic [104] solvents									
Methanol	MeOH	−97.68 [1,2]	64.515 [3]	32.63 [4–7]	0.542 [4,7]	0.7863_6 [8,9]	19.1	55.5	41.5
Ethanol	EtOH	−114.15 [1,10]	78.293 [11]	24.35 [4,12]	1.087 [12,13]	0.7849_5 [8,9]	19.6	51.9	37.9
1-Propanol	PrOH	−126.2 [1,2]	97.151 [11]	20.33_5 [4,14]	1.96_8 [7,15,16]	0.7995 [7,8,9,16]		50.7	37.3
2-Propanol	i-PrOH	−87.90 [10]	82.242 [11]	19.40 [7,17]	2.08 [18,19]	0.7809_7 [9,18,19]		48.6	33.6
1-Butanol	BuOH	−88.62 [1]	117.725 [11]	17.43 [14]	2.61 [15,20]	0.8057_5 [9,20]		50.2	36.8
2-Butanol	i-BuOH	−114.7 [1,2]	99.512 [11]	$16._7$ [1,21]	3.18 [1]; 30 °C	0.8023_9 [9,22]		47.1	
1,2-Ethanediol	EG	−13 [1]	197.3 [1]	40.8 [23]	$16._6$ [24,25,101]	1.109_5 [24,25,101]		56.3	(42.8)
Glycerol		18.18 [1]	290.0 [1]	42.5 [1]	945 [1]	1.2551 [1]; 30 °C			(44.3)
Diethylene glycol	DEG	−6.5 [1]	244.8 [1]	31 [23,26]	17.4 [27]; 30 [1]	1.11_6 [23,27]		53.8	
Triethylene glycol	TEG	−4.3 [1]	288.0 [1]	23.4 [26]	49.0 [1]; 20 °C	1.1235 [1]; 20 °C		53.5	
2-Methoxyethanol	MOE	−85.1 [1]	124.6 [1]	$17._0$ [7,28]	1.53_5 [7,29]	0.9601_4 [7]		52.3	
Cyclohexanol		25.15 [1]	161.1 [1]	15.0 [1]	41.067 [1]	0.9684 [1]		46.9	
Benzylalcohol		−15.3 [1]	205.45 [1]	13.27 [30]	4.650 [1]; 30 °C	1.0424 [31]		50.8	(31.2)
2. Amphiprotic protogenic [103] solvents									
Acetic Acid		16.66 [1]	117.90 [1]	6.20 [34]; 20 °C	1.13_5 [32,33]	1.0436 [1,33]		51.2	52.9
3. Protophilic [103] H-bond donor [104] solvents									
Formamide	FA	2.5_5 [1,38]	218 [38]	$109._5$ [36,37]	3.3_0 [37–39]	1.129_3 [35,38,40,41]	24	56.6	39.8
N-Methylformamide	NMF	−5.4 [38]	180–185 [1,38]	186.9 [42]	1.65 [1,38]	0.998244 [43]		54.1	32.1
N-Methylacetamide	NMA	30.56 [44]	206 [1]	171.7 [45]; 35 °C	3.38_7 [45,46]; 35 °C	0.94553 [46]; 35 °C		52.0	(33.7)
1,2-Diaminoethane	DAE	11.3 [1]	117.26 [1]	1.54 [1]	12.9 [1]	0,8922 [60]	55	42.0	20.9

2-Pyrrolidinone		25 [1]	245 [1]	33 [60]	13.3 [1]	1.107 [1]			
2-Aminoethanol		10.53 [1]	170.95 [1]	37.72 [1]	19.346 [1]	1.01159 [1]		51.8	33.7
Morpholine		− 3.1 [1]	128.94 [1]	7.42 [1]	1.792 [1]; 30 °C	0.99547 [1]			17.5
4. Aprotic protophilic [103] solvents									
Dimethylformamide	DMF	− 60.44 [47]	153.0 [1]	36.71 [1]	0.793_9 [48]	0.9438_7 [40,48]	26.6	43.8	16.0
Diethylformamide	DEF		178.3 [106]	28.4 [42]	1.139 [105]	0.9017 [105,42]	30.9		
Dimethylacetamide	DMA	− 20 [1]	166.1 [1]	37.78 [1]	0.927 [105]	0.9350 [105]	27.8	43.7	13.6
Diethylacetamide	DEA		184 [106]	30.4 [42]	1.226 [105]	0.904 [105,42]	32.3		
1-Methyl-2-pyrrolidinone	NMP	− 24.4 [1]	202 [1]	32.0 [1]	1.663 [105]	1.0286 [105]	27.3	42.2	13.3
Hexamethylphosphoric triamide	HMPT	7.20 [1]	233 [1]	$29._6$ [49–51]	3.22_4 [50,52]	1.0201 [49,52]	38.8	40.9	10.6
Pyridine	PY	− 41.55 [1]	115.256 [1]	12.4 [1]; 21 °C	0.884 [1]	0.97824 [1]	33.1	40.2	14.2
Tetramethylurea	TMU	− 1.2 [1]	175.2 [1]	$23._5$ [53,107]	1.39_8 [53,107]	0.9619 [53,107]	29.6	41.0	(9.8)
Dimethylsulfoxide	DMSO	18.54 [1]	189.0 [1]	$46._5$ [54–57]	1.99_2 [54,56–58]	1.0955 [41,55–57]	29.8	45.0	19.3
5. Aprotic protophobic [103] solvents									
Acetonitrile	AN	− 48.835 [1]	81.60 [1]	35.95 [7,59,61]	0.341 [5,59,61]	0.7767_5 [5,7,59,61]	14.1	46.0	18.9
Propionitrile	PN	− 92.78 [1]	97.35 [1]	27.2 [1]; 20 °C	0.389 [1]; 30 °C	0.77682 [1]	16.1	43.7	
Benzonitrile	BN	− 12.75 [1]	191.10 [1]	25.20 [1]	1.237 [62]	1.0006 [1]	11.9	42.0	15.5
Acetone		− 94.7 [1]	56.29 [1]	20.56 [16,55]	0.303 [16,55,63]	0.7843_5 [16,55,63]	17.0	42.2	12.5
2-Butanone	MEK	− 86.69 [1]	79.64 [1]	18.04_5 [7]	0.368 [7]	0.7996_6 [7]		41.3	
3-Pentanone		− 38.97 [1]	101.99 [1]	17.00 [1]; 20 °C	0.442 [1]	0.80945 [1]		39.3	
Acetophenone		19.62 [1]	202.0 [1]	17.48 [64]	1.511 [1]; 30 °C	1.02382 [1]		41.3	
Nitromethane	NM	− 28.55 [1]	101.20 [1]	38.0 [65]	0.62 [1,65]	1.131 [1,65]	2.7	46.3	20.5
Nitrobenzene	NB	5.76 [1]	210.80 [1]	34.82 [1,66]	1.818 [67]	1.1985 [67,68]	4.4	42.0	14.8
Sulfolane	TMS	28.45 [1]	287.3 [1]	43.3_0 [56,69]; 30 °C	10.28_7 [1,56,69]; 30 °C	1.261_9 [56,69,70]; 30 °C	14.8	44.0	19.2

Table A-I. (continued)

Solvent	Abbreviation	θ_F / °C	θ_B / °C	ε	η / cP	ϱ / g ml^{-1}	DN [110,130] / kcal mol^{-1}	E_T [111] / kcal mol^{-1}	AN [112–114,130]
Ethylene carbonate	EC	36.5 [71]	238 [1]	90.36 [72,73]; 40 °C	1.9 [71,73,74]; 40 °C	1.321_4 [71–74]; 40 °C	16.4		
Propylene carbonate	PC	− 54.53 [75]	242 [60]	64.9_5 [7,76]	2.51_2 [7,77]	1.199_6 [7,77]	15.1	46.6	18.3
γ-Butyrolactone	γ-BL	− 43.53 [1]	204 [1]	39.1 [78]	1.7315 [78]	1.1242 [78]			(18.6)
Acetic anhydride		− 73.1 [1]	140.0 [1]	20 [79]; 30 °C	0.8 [79]; 30 °C	1.0686 [79]; 30 °C	10.5	43.9	
Tributylphosphate	TBP	<−80 [1]	289 [1]	7.959 [1]; 30 °C	3.39 [1]	0.9760 [1]	23.7		9.9
6. Low permittivity-electron donor [104] solvents									
Diethyl ether	DEE	−116.3 [1]	34.55 [1]	4.33 [80]	0.242 [1]; 20 °C	0.70760 [1]	19.2	34.6	3.9
Diisopropyl ether		− 85.5 [1]	68.3 [1]	3.88 [1]	0.379 [1]	0.7182 [1]		34.0	
1,4-Dioxane	DI	11.80 [1]	101.320 [1]	2.21 [1,81]	1.20_4 [82,83]	1.0280_5 [67,82,84]	14.8	36.0	10.8
Tetrahydrofuran	THF	−108.5 [1]	65.965 [85]	7.4_3 [86–88]	0.45_9 [86–88]	0.8819 [89,90]	20.0	37.4	8.0
1,2-Dimethoxyethane	DME	− 58 [1]	84.50 [91]	7.07_5 [92]	0.40_7 [7,92]	0.8612_2 [7,92]		38.2	10.2
Diglyme	DG	− 64 [108]	163 [93]	7.2 [109]	1.06 [94]	0.9440 [1]		38.6	9.9
7. Inert Solvents									
Dichloromethane		− 95.14 [1]	39.75 [1]	8.93 [1]	0.413 [95]	1.31678 [1]			20.4
Trichloroethylene		− 86.4 [1]	87.19 [1]	3.42 [1]; ~16 °C	0.532 [1]	1.4514 [1]; 30 °C		35.9	
Benzene		5.533 [1]	80.0_7 [1,96]	2.27_4 [1,97,98]	0.6028 [1,99]	0.87360 [100]	0.1	34.5	8.2
Cyclohexane		6.544 [1]	80.725 [1]	2.023 [1]	0.898 [1]	0.77374 [102]		31.2	

Table A-II. Radii of ions X_i and distance parameters s

Ion	a_i/nm	Ion	a_i/nm	Ion	a_i/nm	Ion	a_i/nm	Ion	a_i/nm [121]	solvent	s/nm [124]
Li^+	(0.078)	Be^{2+}	(0.034)	Me_4N^+	0.347 [116]	F^-	0.133	CO_3^{2-}	0.178	OH_2	0.28
Na^+	0.098	Mg^{2+}	0.078	$Me_2R_2N^+$	0.347 [116]	Cl^-	0.181	CrO_4^{2-}	0.256	MeOH	0.47
K^+	0.133	Ca^{2+}	0.106	Et_4N^+	0.400 [116]	Br^-	0.196	SO_4^{2-}	0.258	EtOH	0.57
Rb^+	0.149	Sr^{2+}	0.127	Pr_4N^+	0.452 [116]	I^-	0.220	SeO_4^{2-}	0.249	PrOH	0.69
Cs^+	0.165	Ba^{2+}	0.143	n-Bu_4N^+	0.494 [116]	BF_4^-	0.232 [121]	SiF_6^{2-}	0.259	AN	0.58
Cu^+	0.095	Zn^{2+}	0.069	n-Am_4N^+	0.529 [116]	ClO_4^-	0.240 [121]			PC	0.65
Ag^+	0.113	Cd^{2+}	0.103	$AsPh_4^+$	(0.42) [118]	CN^-	0.191 [121]			NM	0.52
NH_4^+	0.137 [121]	Hg^{2+}	0.093			OH^-	0.133 [121]			NB	0.77
		Pb^{2+}	0.117			BPh_4^-	0.48 [124]			(:OH)	0.28
		Mn^{2+}	0.091							|	
		Co^{2+}	0.082								
		Ni^{2+}	0.068								
		Cu^{2+}	0.072								

If more than one reliable value was found in the recent literature the table contains their mean value; the number of decimal places indicates their agreement. In the case that no recent determination of data was found which is based on thoroughly purified solvents and established precision methods we preferred to quote the data from Ref. [1]. The compilation corresponds to the actual state of the ELDAR data bank [115]. It may also be helpful in finding new purification methods, new precise measuring equipment and new precision methods for the determination of data.

A.2 Ion Distance Parameters

The chemical model, Section 4.2., when used to provide data of electrolyte solutions by computer-assisted methods, presupposes the knowledge of the distance parameters a and R. These are needed in the appropriate equations as the characteristic distances of the bare or solvated single ions and of the adjacent solvent molecules. Solvent molecules may be orientated either by free electrons, by H-bonds or by their dipole moments, depending on the nature of the ions and the solvent molecule itself.

Ionic radii depend on the method of their determination. Various systems can be found and are compared in the literature, cf. [116–123] and literature quoted there. Table A-II shows a selection of values which have proved to be compatible with one another [115,124]. Goldschmidt's ionic radii were taken when not indicated otherwise. The distances s due to solvent molecules were calculated from bond lengths, bond angles and van der Waals volumes of functional groups [125,126]. For their use in the chemical model see Section 4.2., cf. also Ref. [124].

B Electrostatic Potential around a Particle with an Arbitrary Charge Distribution

The electrostatic potential around a particle i, ion or molecule, in solution is obtained by the resolution of a system of differential equation, e.g.

$$\nabla^2\psi_i^{(1)}(r, \theta, \varphi) = 0 \qquad a \leqq r \leqq R_1 \qquad \text{(B-1a)}$$

$$\nabla^2\psi_i^{(2)}(r, \theta, \varphi) = 0 \qquad R_1 \leqq r \leqq R_2 \qquad \text{(B-1b)}$$

$$\nabla^2\psi_i^{(3)}(r, \theta, \varphi) = -\frac{\varrho^{(3)}(r, \theta, \varphi)}{\varepsilon_0\varepsilon} \quad . \quad R_2 \leqq r \leqq \infty \qquad \text{(B-1c)}$$

if the space around the appropriate charge distribution of the ion or molecule, situated in a sphere of radius a, is subdivided into three parts:

region (1): $a \leqq r \leqq R_1$, relative permittivity ε_1 and free of single charges (ions), *i.e.* $\varrho^{(1)}(r, \theta, \varphi) = 0$, region (2): $R_1 \leqq r \leqq R_2$, relative permittivity ε_2 and free of single charges (ions), $\varrho^{(2)}(r, \theta, \varphi) = 0$,

region (3): $R_2 \leqq r \leqq \infty$, relative permittivity ε of the bulk solution and charge density $\varrho^{(3)}(r, \theta, \varphi) = -\varepsilon_0 \varepsilon \varkappa^2 \psi_i^{(3)}(r, \theta, \varphi)$ [for $\varkappa$ see table II]

The appropriate boundary conditions for Eqs. (1) are

$$\lim_{R_1 \to \infty} \psi_i^{(1)} = \psi_i^{(0)}(r, \theta, \varphi) \quad \text{(B-2a)} \qquad \psi_i^{(2)}(R_2) = \psi_i^{(3)}(R_2) \quad \text{(B-2d)}$$

$$\psi_i^{(1)}(R_1) = \psi_i^{(2)}(R_1) \quad \text{(B-2b)} \qquad \varepsilon_2 \left(\frac{\partial \psi_i^{(2)}}{\partial r}\right)_{R_2} = \varepsilon \left(\frac{\partial \psi_i^{(3)}}{\partial r}\right)_{R_2} \quad \text{(B-2e)}$$

$$\varepsilon_1 \left(\frac{\partial \psi_i^{(1)}}{\partial r}\right)_{R_1} = \varepsilon_2 \left(\frac{\partial \psi_i^{(2)}}{\partial r}\right)_{R_1} \quad \text{(B-2c)} \qquad \lim_{r \to \infty} \psi_i^{(3)} = 0 \quad \text{(B-2f)}$$

where $\psi_i^{(0)}(r, \theta, \varphi)$ means the potential of the isolated molecular or ionic charge distribution, which is placed in a homogeneous and isotropic medium of relative permittivity ε_1, in terms of Legendre polynomials

$$\psi_i^{(0)} = \sum_{j=0}^{\infty} \sum_{m=-j}^{m=+j} b_j^{(m)} \times \frac{1}{r^{j+1}} \times P_j^{(m)}(\cos\theta)^{im\varphi} \quad \text{(B-3a)}$$

$$b_j^{(m)} = \frac{1}{4\pi\varepsilon_0\varepsilon_1} b_j'^{(m)}. \quad \text{(B-3b)}$$

The coefficients $b_j'^{(m)}$ are calculated from bond lengths, bond angles and charges of the ion or molecule, cf. Ref. [128]. The system of differential equations is resolved by usual methods yielding [127]

$$\psi_i^{(1)}(r, \theta, \varphi) = \sum_{j=0}^{\infty} \sum_{m=-j}^{m=+j} \left[A_j^{(m)} \times r^j + B_j^{(m)} \times \frac{1}{r^{j+1}}\right] \times P_j^{(m)}(\cos\theta) \times e^{im\varphi} \quad \text{(B-4a)}$$

$$\psi_i^{(2)}(r, \theta, \varphi) = \sum_{j=0}^{\infty} \sum_{m=-j}^{m=+j} \left[C_j^{(m)} \times r^j + D_j^{(m)} \times \frac{1}{r^{j+1}}\right] \times P_j^{(m)}(\cos\theta) \times e^{im\varphi} \quad \text{(B-4b)}$$

$$\psi_i^{(3)}(r, \theta, \varphi) = \sum_{j=0}^{\infty} \sum_{m=-j}^{m=+j} \left[E_j^{(m)} \times e^{\varkappa r} \sum_{n=0}^{j} \frac{\alpha_n^{(j)}}{r^{n+1}} + F_j^{(m)} \times e^{-\varkappa r} \sum_{n=0}^{j} \frac{\beta_n^{(j)}}{r^{n+1}}\right] \times \times P_j^{(m)}(\cos\theta) \times e^{im\varphi} \quad \text{(B-4c)}$$

with the coefficients $\alpha_n^{(j)}$ and $\beta_n^{(j)}$ of Eq. (4c) obtained from the recursion formulae

$$[n(n+1) - j(j+1)]\,\alpha_n^{(j)} - 2\varkappa(n+1)\,\alpha_{n+1}^{(j)} = 0; \qquad \alpha_0^{(j)} = 1 \quad \text{(B-5a)}$$

$$[n(n+1) - j(j+1)]\,\beta_n^{(j)} + 2\varkappa(n+1)\,\beta_{n+1}^{(j)} = 0; \qquad \beta_0^{(j)} = 1 \quad \text{(B-5b)}$$

Combining Eqs. (3), (4) and (5) gives

$$A_j^{(m)} = -\frac{(j+1)(1-\eta)X_j - [j(1+\eta)+\eta]Q^{2j+1}Y_j}{[j(1+\eta)+1]X_j - j(1-\eta)Q^{2j+1}Y_j} \times \frac{1}{R_1^{2j+1}} \times b_j^{(m)} \tag{B-6a}$$

$$B_j^{(m)} = b_j^{(m)} \tag{B-6b}$$

$$C_j^{(m)} = \frac{\eta(2j+1)Y_j}{[j(1+\eta)+1]X_j - j(1-\eta)Q^{2j+1}Y_j} \times \frac{1}{R_2^{2j+1}} \times b_j^{(m)} \tag{B-6c}$$

$$D_j^{(m)} = \frac{\eta(2j+1)X_j}{[j(1+\eta)+1]X_j - j(1-\eta)Q^{2j+1}Y_j} \times b_j^{(m)} \tag{B-6d}$$

$$E_j^{(m)} = 0 \tag{B-6e}$$

$$F_j^{(m)} = \frac{\eta(2j+1)^2 \times \exp(\varkappa R_2)}{[j(1+\eta)+1]X_j - j(1-\eta)Q^{2j+1}Y_j} \times b_j^{(m)} \tag{B-6f}$$

with $\eta = (\varepsilon_1/\varepsilon_2)$ and $Q = (R_1/R_2)$. The X_j and Y_j are polynomials

$$X_j = \sum_{n=0}^{j}\left[\frac{\varepsilon}{\varepsilon_2}(1+n+\varkappa R_2)+j\right]R_2^{j-n}\beta_n^{(j)} \tag{B-7a}$$

$$Y_j = -\sum_{n=0}^{j}\left[\frac{\varepsilon}{\varepsilon_2}(1+n+\varkappa R_2)-(j+1)\right]R_2^{j-n}\beta_n^{(j)} \tag{B-7b}$$

A tabular survey for j = 0 to j = 3 is given in tables B-I and B-II

Table B-I: Polynomials X_j

$$\varepsilon_2 X_0 = \varepsilon(1+\varkappa R_2)$$
$$\varkappa\varepsilon_2 X_1 = \varepsilon(2+2\varkappa R_2+\varkappa^2R_2^2)+\varepsilon_2(1+\varkappa R_2)$$
$$\varkappa^2\varepsilon_2 X_2 = \varepsilon(9+9\varkappa R_2+4\varkappa^2R_2^2+\varkappa^3R_2^3)+2\varepsilon_2(3+3\varkappa R_2+\varkappa^2R_2^2)$$

Table B-II: Polynomials Y_j

$$\varepsilon_2 Y_0 = -\varepsilon(1+\varkappa R_2)+\varepsilon_2$$
$$\varkappa\varepsilon_2 Y_1 = -\varepsilon(2+2\varkappa R_2+\varkappa^2R_2^2)+2\varepsilon_2(1+\varkappa R_2)$$
$$\varkappa^2\varepsilon_2 Y_2 = -\varepsilon(9+9\varkappa R_2+4\varkappa^2R_2^2+\varkappa^3R_2^3)+3\varepsilon_2(3+3\varkappa R_2+\varkappa^2R_2^2)$$

The electrostatic parts $W_{ij}^{el}(r)$ of the ion-ion interaction potentials in Table II

$$W_{ij}^{el} = e_0 z_j \psi_i^{(2)} \qquad (a \leqq r \leqq R) \quad \text{and} \quad W_{ij}^{el} = e_0 z_j \psi_i^{(3)} \qquad (r \geqq R)$$

are obtained by setting
$b_j'^{(m)} = b_0'^{(0)} = e_0 z_i$ (spherical charge symmetry of the ion i). Local permittivities are not considered ($\varepsilon_1 = \varepsilon_2 = \varepsilon$) and $R_1 = R_2 = R$, *i.e.* $\psi_i^{(1)} = \psi_i^{(2)}$.

The kinetic solvent effect, Section VIII, presupposes vanishing ional concentrations ($R_2 \to \infty$; $\varepsilon_2 = \varepsilon$). The parameter R_1 is the distance of the reactants, i and j, in the activated complex ($R_1 = r_0$); $\varepsilon_1 = 1$ for a non-polarisable particle. The use of first order approximations, *i.e.* single charge ($b_j'^{(m)} = b_0'^{(0)} = e_0 z_i$) or point dipole ($b_j'^{(m)} = b_1'^{(0)} = \mu_i$), for the ion or polar molecule, respectively, yields the Equations (51) and (52) of Section 8.2.

C Electrostatic Potential around a Polarisable Apolar Particle

A polarisable apolar molecule can be represented by a dielectric sphere of radius a and relative permittivity ε_i bearing no charge distribution which can produce singularities for $r \leqq a$. An ion or polar molecule situated in the vicinity of this sphere give rise to an induced dipole moment in and hence to an electrostatic field around the dielectric sphere. The solvent around the apolar molecule is considered as a homogeneous medium of relative permittivity ε.

The appropriate potential equations are

$$\psi_i^{(1)} = \sum_{j=0}^{\infty} \sum_{m=-j}^{m=+j} \left[A_j^{(m)} \times r^j + B_j^{(m)} \times \frac{1}{r^{j+1}} \right] \times P_j^{(m)}(\cos\theta) \times e^{im\varphi} \qquad r \leqq a \tag{C-1 a}$$

$$\psi_i^{(2)} = \sum_{j=0}^{\infty} \sum_{m=-j}^{m=+j} \left[C_j^{(m)} \times r^j + D_j^{(m)} \times \frac{1}{r^{j+1}} \right] \times P_j^{(m)}(\cos\theta) \times e^{im\varphi} \qquad r \geqq a. \tag{C-1 b}$$

with the boundary conditions [129]

$$\lim_{r\to 0} \psi_i^{(1)} = \text{finite} \tag{C-2 a}$$

$$\varepsilon_i \left(\frac{\partial \psi_i^{(1)}}{\partial r} \right)_a = \varepsilon \left(\frac{\partial \psi_i^{(2)}}{\partial r} \right)_a \tag{C-2 c}$$

$$\psi_i^{(1)}(a) = \psi_i^{(2)}(a) \tag{C-2 b}$$

$$\lim_{r\to\infty} \varphi_i^{(2)} = 0 \tag{C-2 d}$$

with

$$\varphi_i^{(2)}(P) = \psi_i^{(2)}(P) - \psi^{(0)}(P) \tag{C-3 a}$$

$$\psi^{(0)}(P) = \sum_{j=0}^{\infty} \sum_{m=-j}^{m=+j} C_j^{0(m)} \times r^j \times P_j^{(m)}(\cos\theta) \times e^{im\varphi} \tag{C-3 b}$$

$\psi^{(0)}(P)$ is that potential which would exist at point $P(r, \theta, \varphi)$ if the dielectric sphere of relative permittivity ε_i would be replaced by the dielectric medium of relative permittivity ε.

A single charge (ion) situated at a distance d from the center of the dielectric sphere yields a potential

$$\psi^{(0)}(P) = \frac{e_0}{4\pi\varepsilon_0\varepsilon} \sum_{j=0}^{\infty} \frac{r^j}{d^{j+1}} P_j(\cos\theta) \tag{C-4}$$

This special case is the basic model of Section 7.2. for the reaction between halides and olefines. The appropriate potentials are

$$\psi_i^{(1)} = \frac{e_0}{4\pi\varepsilon_0\varepsilon} \sum_{j=0}^{\infty} \frac{(2j+1)\,\varepsilon}{(j+1)\,\varepsilon + j\varepsilon_i} \frac{r^j}{d^{j+1}} P_j(\cos\theta) \qquad r \leqq a \tag{C-5 a}$$

$$\psi_i^{(2)} = \frac{e_0}{4\pi\varepsilon_0\varepsilon} \sum_{j=0}^{\infty} \left[\frac{r^j}{d^{j+1}} + \frac{(\varepsilon - \varepsilon_i)\, j}{(j+1)\,\varepsilon + j\varepsilon_i} \times \frac{a^{2j+1}}{(r \times d)^{j+1}}\right] P_j(\cos\theta) \qquad r \geqq a \tag{C-5 b}$$

From Eqs. (C-4) and (C-5) the following field and dielectric displacement vectors are obtained

$$\vec{E}_1^{(0)} = -\vec{\text{grad}}\, \psi^{(0)}; \qquad \vec{D}_1^{(0)} = \varepsilon_0\varepsilon\vec{E}_1^{(0)} \qquad 0 \leqq r \leqq \infty \tag{C-6}$$

and

$$\vec{E}_2^{(1)} = -\vec{\text{grad}}\, \psi_i^{(1)}; \qquad \vec{D}_2^{(1)} = \varepsilon_0\varepsilon_i\vec{E}_2^{(1)} \qquad 0 \leqq r \leqq a \tag{C-7 a}$$

$$\vec{E}_2^{(2)} = -\vec{\text{grad}}\, \psi_i^{(2)}; \qquad \vec{D}_2^{(2)} = \varepsilon_0\varepsilon\vec{E}_2^{(2)} \qquad a \leqq r \leqq \infty \tag{C-7 b}$$

Eq. (C-6) represents the situation before and Eqs. (C-7) after the dielectric sphere is brought from infinite distance to a distance d from the ion. Hence, the energy for building the latter configuration is

$$\Delta G = \frac{N_A}{2} \iiint (\vec{E}_2\vec{D}_2 - \vec{E}_1\vec{D}_1)\, dV \tag{C-8}$$

Eq. (C-8) can be transformed [129] to give

$$\begin{aligned}\Delta G &= N_A \frac{\varepsilon - \varepsilon_i}{2} \iiint \vec{E}_1\vec{E}_2\, dV \\ &= N_A e_0^2 \times \frac{\varepsilon - \varepsilon_i}{8\pi\varepsilon_0\varepsilon} \times \frac{1}{d} \sum_{j=0}^{\infty} \frac{j}{(j+1)\,\varepsilon + j\varepsilon_i} \left(\frac{a}{d}\right)^{2j+1}.\end{aligned} \tag{C-9}$$

Eq. (C-9) is an ill converging series if $d \approx a$ as it is required for building the activated complex of the reaction between an apolar molecule and an ion.

D Symbols and Abbreviations

D.1 Fundamental Constants [131]

N_A: Avogadro constant	$(6.022045 \pm 0.000031) \times 10^{23}$ mol^{-1}
F: Faraday constant	$(9.648456 \pm 0.000027) \times 10^{4}$ C mol^{-1}
h: Planck constant	$(6.626176 \pm 0.000036) \times 10^{-34}$ J s
k: Boltzmann constant	$(1.380662 \pm 0.000044) \times 10^{-23}$ J K^{-1}
e_0: charge of proton	$(1.602189 \pm 0.000005) \times 10^{-19}$ C
ε_0: permittivity of vacuum	$(8.854188 \pm 0.000000) \times 10^{-12}$ J^{-1} C^{2} m^{-1}

D.2 Generally Used Superscripts and Subcripts

D.2.1 Superscripts

*: pure substance also: non-Coulombic part of a quantity
$^\infty$: infinite dilution
^{0}: standard
$^{(0)}$: reference solvent (Section VIII)
$^{(s)}$: solvent S (Section VIII)
‡: activated complex
^{el}: Coulombic part of a quantity
$'$ free particles also: final state (Section V)
$^{(c)}$: molar scale
$^{(m)}$: molal scale
$^{(i)}$: component i

D.2.2 Subscripts

$_A$: association
$_{calc}$: calculated quantity
$_{exp}$: experimental quantity
$_{FI}$: free ion
$_i$or$_j$: ions i or j also: index for summation formulae $\sum_i$
$_{ij}$: ion j in the vicinity of an ion i
$_{IP}$: ion pair
$_{max}$: maximum value
$_{obs}$: observed quantity
$_S$: solvent
$_Y$: electrolyte Y
$_+$: cation
$_-$: anion
$_\pm$: mean quantity of electrolyte Y
$_\infty$: infinite frequency

in connexion with symbol Δ, cf. Fig. 1

$_{dil}$: dilution
$_{lat}$: lattice
$_{sol}$: solution
$_{solv}$: solvation
$_{vap}$: vaporation
$_{tr}$: transfer from water to a non-aqueous solvent

D.3 Symbols

* means a superscript or subscript, see Section D.2.

a, a_*:	ion size parameter
c, c_*, c^*:	molarity [mol dm^{-3}]
E, E^*:	property (general) of an electrolyte solution (Sect. V)
	also: e.m.f (Sect. V)
f, f_S:	activity coefficient (mole fraction scale)
f_{ij}, f_{ji}:	two-particle distribution function
g_{ij}, g_{ji}:	pair-correlation function
$\Delta G_*, \Delta G^*, \Delta_* G$:	Gibbs energy
$\Delta H_*, \Delta H^*, \Delta_* H$:	enthalpy
K, K_*:	equilibrium constant
k, k_*, k^*:	rate constant, exception see D.1
m, m_*, m^*:	molality [mol kg^{-1}]
	also: mass of particle (Sect. IV)
M:	mol dm^{-3}
n_S, n_Y:	amount of substance (number of moles)
N_*:	particle density [particles per cm^{-3}], exception see D.1
p:	pressure
q:	Bjerrum parameter, see Eq. (19b)
R, R_*:	distance parameter, see Figs. 2 and 3
R	also: gas constant
$r, r_*, \vec{r}, \vec{r}_*$:	distance variable
$\Delta S^*, \Delta S_*, \Delta_* S$:	entropy
t_*, t^*:	transference number
t	also: time
T, T_*:	temperature [K]
$V, \Delta V^*, \Delta V_*$:	volume
W_*, W^*:	mean-force potential
x, x_*, x^*:	mole fraction
y, y_*, y^*:	activity coefficient (molar scale)
z, z_*:	valency
$Z, \dot{Z}_*, Z^*, \Delta Z_*, \Delta Z^*, \Delta_* Z$:	thermodynamic property (general), see Fig. 1
α:	degree of dissociation
$(1-\alpha)$:	degree of association
$\bar{\alpha}$:	cubic expansion coefficient (Section V)
γ_*, γ^*:	activity coefficient (molal scale)
${}_m\gamma_*$:	transfer activity coefficient
Γ:	ional concentration, see Table II
δ, δ_*:	chemical shift
$\varepsilon, \varepsilon_*, \varepsilon^*$:	relative permittivity, exception see D.1
η:	viscosity
θ, θ_*:	temperature [°C]
$\varkappa$:	Debye parameter, see Table II
$\varkappa, \varkappa_*$:	also: electrolytic conductivity [Ω^{-1} cm^{-1}]
Λ, Λ^*:	equivalent conductance [Ω^{-1} mol^{-1} cm^2]

λ_*, λ^*:	single ion conductance [Ω^{-1} mol^{-1} cm^2]
μ_*, μ^*:	chemical potential
$\nu_i \nu_i^m$:	stoichiometric number
ϱ:	density [g cm^{-3}]
ω:	circular frequency

Remark: Symbols used only at single places with special meaning, which may also deviate from the meaning of the above list, are explained in the text.

XIX References (Part A and B)

1. Parker, A. J.: Electrochim. Acta *21*, 671 (1976)
2. Friedman, H. L.: Faraday Discuss. Chem. Soc. *64*, 7 (1977)
3. Barthel, J.: Ber. Bunsenges. Phys. Chem. *83*, 252 (1979)
4. Barthel, J.: Pure Appl. Chem. *51*, 2093 (1979)
5. Bjerrum, N.: Math.-Fys. Medd. Kon. Dan. Vidensk. Selsk. *7*, 1 (1926)
6. Kosower, E. M.: J. Amer. Chem. Soc. *80*, 3253 (1958)
7. Kosower, E. M.: J. Chim. Phys. Phys.-Chim. Biol. *61*, 230 (1964)
8. Dimroth, K., Reichardt, C., Siepmann, T. and Bohlmann, F.: Liebigs Ann. Chem. *661*, 1 (1963)
9. Reichardt, C.: Solvent Effects in Organic Chemistry, Weinheim: Verlag Chemie 1979
10. Gutmann, V. and Wychera, E.: Inorg. Nucl. Chem. Lett. *2*, 257 (1966)
11. Gutmann, V.: The Donor-Acceptor Approach to Molecular Interactions, New York: Plenum 1978
12. Grunwald, E. and Winstein, S.: J. Amer. Chem. Soc. *70*, 846 (1948)
13. Gielen, M. and Nasielsky, J.: J. Organomet. Chem. *1*, 173 (1963)
14. Gielen, M. and Nasielsky, J.: J. Organomet. Chem. *7*, 273 (1967)
15. Mayer, U., Gutmann, V. and Gerger, W.: Monatsh. Chem. *106*, 1235 (1975)
16. Mayer, U.: Coord. Chem. Rev. *21*, 159 (1976)
17. Berson, J. A., Hamlet, Z. and Mueller, W. A.: J. Amer. Chem. Soc. *84*, 297 (1962)
18. Reichardt, C.: Angew. Chem. *77*, 30 (1965)
19. Burgess, J.: Metal Ions in Solution, Chichester: Ellis Horwood 1978
20. Dack, M. R. J. in: Dack, M. R. J. (ed.): Techniques of Chemistry, Vol. 8, Solutions and Solubilities (Part II), p. 95, New York: Wiley 1976
21. Jensen, W. B.: The Lewis Acid-Base Concepts, New York: Wiley 1980
22. Debye, P. and Hückel, E.: Phys. Z. *24*, 185 (1923)
23. Pimentel, G. C. and McClellan, A. L.: The Hydrogen Bond, San Francisco: Freeman 1960
24. Craver, J. K.: J. Appl. Polym. Sci. *14*, 1755 (1970)
25. Godfrey, N. B.: Chem. Techn. *1972*, 359
26. Kolthoff, I. M. in: Marchon, J. C. (ed.): Non Aqueous Electrochemistry, p. 312, London: Butterworths 1971
27. Bates, R. G. in: Coetzee, J. F. and Ritchie, C. D. (eds.): Solute-Solvent Interactions, Vol. 1, p. 45, New York: Dekker 1969
28. Kratochvil, B.: Rec. Chem. Prog. *27*, 253 (1966)
29. Brönsted, J. N.: Chem. Ber. *61*, 2049 (1928)
30. Fuoss, R. M. and Accascina, F.: Electrolytic Conductance, New York: Interscience 1959
31. Kortüm, G.: Lehrbuch der Elektrochemie (4th ed.) Weinheim: Verlag Chemie 1966

32. Kolthoff, I. M. and Bruckenstein, S.: J. Amer. Chem. Soc. *78*, 1 (1956)
33. Bruckenstein, S. and Kolthoff, I. M.: J. Amer. Chem. Soc. *78*, 10 (1956)
34. Bruckenstein, S. and Kolthoff, I. M.: J. Amer. Chem. Soc. *78*, 2974 (1956)
35. Hantzsch, A.: Z. Phys. Chem. *A134*, 406 (1928)
36. Barthel, J.: Ionen in nichtwäßrigen Lösungen, Darmstadt: Steinkopff 1976
37. Staveley, L. A. K. (ed.): The Characterization of Chemical Purity. Organic Compounds, London: Butterworths 1971
38. Zief, M. (ed.): Purification of Inorganic and Organic Materials, New York: Dekker 1969
39. Zief, M. and Speights, R. (eds.): Ultrapurity Methods and Techniques, New York: Dekker 1972
40. Perrin, D. D., Armarego, W. L. F. and Perrin, D. R.: Purification of Laboratory Chemicals (2nd ed), Oxford: Pergamon 1980
41. Mann, C. K. in: Bard, A. J. (ed.): Electroanalytical Chemistry, Vol. 3, New York: Dekker 1969
42. Riddick, J. A. and Bunger, W. B. in: Weissberger, A. (ed.): Techniques of Chemistry, Vol. 2 (3rd ed.): Organic Solvents, New York: Wiley 1970
43. Kirkwood, J. G. and Oppenheim, I.: Chemical Thermodynamics, New York: McGraw-Hill 1961
44. Robinson, R. A. and Stokes, R. H.: Electrolyte Solutions (2nd ed.), London: Butterworths 1970
45. Onsager, L.: Phys. Z. *28*, 277 (1927)
46. Falkenhagen, H.: Theorie der Elektrolyte (2nd ed.), Stuttgart: S. Hirzel 1971
47. Harned, H. S. and Owen, B. B.: The Physical Chemistry of Electrolytic Solutions (3rd ed.), New York: Reinhold 1958
48. Résibois, P. M. V.: Electrolyte Theory, New York: Harper and Row 1968
49. Friedman, H. L.: Ionic Solution Theory, New York: Interscience 1962
50. Petrucci, S. (ed.): Ionic Interactions, Vol. 1, Vol. 2, New York: Academic Press 1971
51. Kelbg, G.: Z. Phys. Chem. (Leipzig) *214*, 8 (1960)
52. Falkenhagen, H. and Ebeling, W. in: Ref. [50], Vol. 1, p. 1; Falkenhagen, H., Ebeling, W. and Kraeft, W. D. in: Ref. [50], Vol. 1, p. 61
53. Falkenhagen, H. and Kelbg, G. in: Bockris, J. O'M. (ed.): Modern Aspects of Electrochemistry, Vol. 2, p. 1, London: Butterworths 1959
54. Conway, B. E. and Barradas, R. G. (eds.): Chemical Physics of Ionic Solutions, New York: Wiley 1966
55. Covington, A. K. and Dickinson, T. (eds.): Physical Chemistry of Organic Solvent Systems, London: Plenum 1973
56. Rasaiah, J. C. and Friedman, H. L.: J. Phys. Chem. *72*, 3352 (1968); J. Chem. Phys. *50*, 3965 (1969)
57. Rasaiah, J. C.: J. Chem. Phys. *52*, 704 (1970); *56*, 3071 (1972)
58. Rasaiah, J. C.: J. Solution Chem. *2*, 301 (1973)
59. Friedman, H. L. and Larsen, B.: J. Chem. Phys. *70*, 92 (1979)
60. Friedman, H. L.: Annu. Rev. Phys. Chem. *32*, (1981)
61. Friedman, H. L.: Pure Appl. Chem. *53*, 1277 (1981)
62. Fuoss, R. M.: Proc. Natl. Acad. Sci. USA *75*, 16 (1978)
63. Ebeling, W. and Grigo, M.: Ann. Phys. (Leipzig) *36*, 21 (1980)
64. Ebeling, W. and Künstner, H. (eds.): Theorie elektrolytischer Flüssigkeiten, Rostocker Physikalische Manuskripte, Heft 4, 1979
65. Justice, M.-C. and Justice, J.-C.: J. Solution Chem. *5*, 543 (1976); *6*, 819 (1977)
66. Justice, J.-C.: J. Solution Chem. *7*, 859 (1978)
67. Justice, J.-C. and Ebeling, W.: J. Solution Chem. *8*, 809 (1979)
68. Justice, J.-C. and Justice, M.-C.: Pure Appl. Chem. *51*, 1681 (1979)
69. Justice, J.-C., Justice, M.-C. and Micheletti, C.: Pure Appl. Chem. *53*, 1291 (1981)
70. Hertz, H. G.: Electrochemistry. A Reformulation of the Basic Principles (Lecture Notes in Chemistry, Vol. 17) Heidelberg: Springer 1980
71. European Federation of Chemical Engineering (ed.): Phase Equilibria and Fluid Properties in the Chemical Industry, Vol. 2, Frankfurt: DECHEMA 1980
72. Barthel, J.: in Ref. [71], p. 497
73. Onsager, L.: quoted after Ref. [52]

74. Barthel, J.: Chem.-Ing.-Techn. *50*, 259 (1978)
75. Barthel, J., Wachter, R. and Gores, H.-J.: Faraday Discuss. Chem. Soc. *64*, 285 (1977)
76. Barthel, J., Wachter, R. and Gores, H.-J. in: Conway, B. E. and Bockris, J. O'M. (eds.): Modern Aspects of Electrochemistry, Vol. 13, p. 1, New York: Plenum 1979
77. Wachter, R. and Riederer, K.: Pure Appl. Chem. *53*, 1301 (1981)
78. Barthel, J. and Schmeer, G. in: Bak, T. (ed.): Kemisk Forening 100 Års Jubilaeum: Kopenhagen (in press)
79. Wachter, R. and Barthel, J.: Ber. Bunsenges. Phys. Chem. *83*, 634 (1979)
80. Barthel, J., Schmeer, G., Strasser, F. and Wachter, R.: Rev. Chim. Minérale *15*, 99 (1978)
81. Schwitzgebel, G., Lührs, C. and Barthel, J.: Ber. Bunsenges. Phys. Chem. *84*, 1220 (1980)
82. Barthel, J., Ströder, U., Iberl, L. and Hammer, H.: Ber. Bunsenges. Phys. Chem. *86*, 636 (1982)
83. Barthel, J., Popp, H. and Schmeer, G.: Data Bank ELDAR (*El*ectrolyte *Da*ta Bank *R*egensburg): DECHEMA project
84. Wachter, R. et al.: in preparation
85. Abraham, M. H., Danil de Namor, A. F. and Schulz, R. A.: J. Solution Chem. *6*, 491 (1977)
86. Abraham, M. H., Danil de Namor, A. F. and Schulz, R. A.: J. Solution Chem. *5*, 529 (1976)
87. Abraham, M. H.: J. Chem. Soc., Faraday Trans. 1 *69*, 1375 (1973)
88. Wölbl, J.: Dissertation, Regensburg (1982)
89. Pasztor, A. J. and Criss, C. M.: J. Solution Chem. *7*, 27 (1978)
90. Choi, Y.-S. and Criss, C. M.: Faraday Discuss. Chem. Soc. *64*, 204 (1977)
91. Dack, M. R. J., Bird, K. J. and Parker, A. J.: Aust. J. Chem. *28*, 955 (1975)
92. Millero, F. J.: J. Phys. Chem. *72*, 3209 (1968)
93. Millero, F. J.: J. Phys. Chem. *73*, 2417 (1969)
94. Gopal, R., Agarwal, D. K. and Kumar, R.: Bull. Chem. Soc. Japan *46*, 1973 (1973)
95. Zana, R., Lage, G. A. and Criss, C. M.: J. Solution Chem. *9*, 667 (1980)
96. Smid, J. and Grotens, A. M.: J. Phys. Chem. *77*, 2377 (1973)
97. Chang, S. and Criss, C. M.: J. Solution Chem. *2*, 457 (1973)
98. Libuś, W., Meçik, M. and Strzelecki, H.: J. Solution Chem. *9*, 723 (1980)
99. Choi, Y.-S. and Criss, C. M.: J. Chem. Eng. Data *22*, 297 (1977)
100. Shin, C. and Criss, C. M.: J. Solution Chem. *7*, 205 (1978)
101. Bennetto, H. P. and Spitzer, J. J.: J. Chem. Soc., Faraday Trans. 1 *73*, 1066 (1977)
102. Bennetto, H. P. and Spitzer, J. J.: J. Chem. Soc., Faraday Trans. 1 *72*, 2108 (1976)
103. Spitzer, J. J.: J. Chem. Soc., Faraday Trans. 1 *74*, 756 (1978)
104. Pitzer, K. S.: Acc. Chem. Res. *10*, 371 (1977)
105. Picker, P., Jolicoeur, C. and Desnoyers, J. E.: J. Chem. Thermodyn. *1*, 469 (1969)
106. Wadsö, I.: Acta Chem. Scand. *22*, 927 (1968)
107. Monk, P. and Wadsö, I.: Acta Chem. Scand. *22*, 1842 (1968)
108. Born, M. and Mayer, J. E.: Z. Phys. *75*, 1 (1932)
109. Ladd, M. F. C. and Lee, W. H.: in Reiss, H. (ed.): Progress in Solid State Chemistry, Vol. 1, p. 37, Oxford: Pergamon 1964
110. Morris, D. F. C.: Struct. Bonding (Berlin) *4*, 63 (1968)
111. Criss, C. M. and Salomon, M.: in Ref. [55], p. 253
112. Krishnan, C.V. and Friedman, H. L. in: Coetzee, J. F. and Ritchie, C. D. (eds.): Solute-Solvent Interactions, Vol. 2, p. 1, New York: Dekker 1976
113. Janz, G. J. and Tomkins, R. P. T.: Nonaqueous Electrolytes Handbook, Vol. 2, New York: Academic Press 1973
114. Popovych, O. in: Kolthoff, I. M. and Elving, P. J. (eds.): Treatise on Analytical Chemistry (2nd ed.), Vol. 1 (1), p. 711, New York: Wiley 1978
115. Trémillon, B. and Coetzee, J. F.: Pure Appl. Chem. *50*, 587 (1978)
116. Popovych, O.: Anal. Chem. *38*, 558 (1966)
117. Ben-Naim, A.: J. Phys. Chem. *82*, 792 (1978)
118. Kim, J. I., Duschner, H., Born, H.-J. and Hashimoto, T.: Z. Phys. Chem. N.F. *103*, 15 (1976)
119. Marcus, Y.: Rev. Anal. Chem. *5*, 53 (1980)
120. Coetzee, J. F. and Istone, W. K.: Anal. Chem. *52*, 53 (1980)

121. Coetzee, J. F. and Martin, M. W.: Anal. Chem. *52*, 2412 (1980)
122. Born, M.: Z. Phys. *1*, 45 (1920)
123. Hepler, L. G.: Aust. J. Chem. *17*, 587 (1964)
124. Stiles, P. J.: Aust. J. Chem. *33*, 1389 (1980)
125. Latimer, W. M., Pitzer, K. S. and Slansky, C. M.: J. Chem. Phys. *7*, 108 (1939)
126. Stokes, R. H.: J. Amer. Chem. Soc. *86*, 979 (1964)
127. Dogonadze, R. R. and Kornyshev, A. A.: J. Chem. Soc., Faraday Trans. 2 *70*, 1121 (1974)
128. Bernal, J. D. and Fowler, R. H.: J. Chem. Phys. *1*, 515 (1933)
129. Frank, H. S. and Wen, W.-Y.: Faraday Discuss. Chem. Soc. *24*, 133 (1957)
130. Nemethy, G. and Scheraga, H. A.: J. Chem. Phys. *36*, 3382 (1962)
131. Cox, B. G., Hedwig, G. R., Parker, A. J. and Watts, D. W.: Aust. J. Chem. *27*, 477 (1974)
132. Covington, A. K. and Newman, K. E.: Pure Appl. Chem. *51*, 2041 (1979)
133. Schneider, H.: Topics in Current Chemistry *68*, 103 (1976)
134. Kim, J. I.: Z. Phys. Chem. N.F. *121*, 1 (1980)
135. Kim, J. I. and Duschner, H.: J. Inorg. Nucl. Chem. *39*, 471 (1977)
136. Kim, J. I.: J. Phys. Chem. *82*, 191 (1978)
137. de Valera, E., Feakins, D. and Waghorne, W. E.: J. Chem. Soc., Faraday Trans. 1 *76*, 560 (1980)
138. Buckingham, A. D.: Faraday Discuss. Chem. Soc. *24*, 151 (1957)
139. Muirhead-Gould, J. S. and Laidler, K. J.: Trans. Faraday Soc. *63*, 944 (1967)
140. Morf, W. E. and Simon, W.: Helv. Chim. Acta *54*, 794 (1971)
141. Izmailov, N. A.: Dokl. Akad. Nauk SSSR *126*, 1033 (1959)
142. Izmailov, N. A.: Zh. Fiz. Khim. *34*, 2414 (1960)
143. Feakins, D. and Watson, P.: J. Chem. Soc. *1963*, 4734
144. Alfenaar, M. and De Ligny, C. L.: Recl. Trav. Chim. Pays-Bas *86*, 929 (1967)
145. De Ligny, C. L. and Alfenaar, M.: Recl. Trav. Chim. Pays-Bas *84*, 81 (1965)
146. De Ligny, C. L., Denessen, H. J. M. and Alfenaar, M.: Recl. Trav. Chim. Pays-Bas *90*, 1265 (1971)
147. Pleskov, V. A.: Usp. Khim. *16*, 254 (1947)
148. Köpp, H.-M., Wendt, H. and Strehlow, H.: Z. Electrochem. *64*, 483 (1960)
149. Parker, A. J. and Alexander, R.: J. Amer. Chem. Soc. *90*, 3313 (1968)
150. Alexander, R. and Parker, A. J.: J. Amer. Chem. Soc. *89*, 5549 (1967)
151. Grunwald, E., Baughman, G. and Kohnstam, G.: J. Amer. Chem. Soc. *82*, 5801 (1960)
152. Fuchs, R., Bear, J. L. and Rodewald, R. F.: J. Amer. Chem. Soc. *91*, 5797 (1969)
153. Kolthoff, I. M. and Chantooni, M. K.: J. Phys. Chem. *76*, 2024 (1972)
154. Millero, F. J.: J. Phys. Chem. *75*, 280 (1971)
155. Kim, J. I., Cecal, A., Born, H.-J. and Gomaa, E. A.: Z. Phys. Chem. N.F. *110*, 209 (1978)
156. Kim, J. I.: Z. Phys. Chem. N.F. *113*, 129 (1978)
157. Coetzee, J. F. and Sharpe, W. R.: J. Phys. Chem. *75*, 3141 (1971)
158. Treiner, C.: Can. J. Chem. *55*, 682 (1977)
159. Pierotti, R. A.: Chem. Rev. *76*, 717 (1976)
160. Parker, A. J.: Electrochim. Acta *21*, 671 (1976)
161. Alexander, R., Parker, A. J., Sharp, J. H. and Waghorne, W. E.: J. Amer. Chem. Soc. *94*, 1148 (1972)
162. Owensby, D. A., Parker, A. J. and Diggle, J. W.: J. Amer. Chem. Soc. *96*, 2682 (1974)
163. Lam, S. Y., Louis, C. and Benoit, R. L.: J. Amer. Chem. Soc. *98*, 1156 (1976)
164. Mayer, U.: Monatsh. Chem. *108*, 1479 (1977)
165. Marcus, Y., Pross, E. and Hormadaly, J.: J. Phys. Chem. *84*, 2708 (1980)
166. Pitzer, K. S.: J. Phys. Chem. *77*, 268 (1973)
167. Pitzer, K. S. and Mayorga, G.: J. Phys. Chem. *77*, 2300 (1973)
168. Pitzer, K. S. and Mayorga, G.: J. Solution Chem. *3*, 539 (1974)
169. Pitzer, K. S. and Kim, J. J.: J. Amer. Chem. Soc. *96*, 5701 (1974)
170. Pitzer, K. S., Roy, R. N. and Silvester, L. F.: J. Amer. Chem. Soc. *99*, 4930 (1977)
171. Pitzer, K. S. in: Pytkowicz, R. M. (ed.): Activity Coefficients in Electrolyte Solutions, Vol. 1, p. 157, Boca Raton: CRC Press 1979
172. Cruz, J. L. and Renon, H.: AIChE J. *24*, 817 (1978)
173. Cruz, J. L. and Renon, H.: Ind. Eng. Chem. Fundament. *18*, 168 (1979)
174. Renon, H.: in Ref. [71], p. 509

175. Smedly, S. I.: The Interpretation of Ionic Conductivity in Liquids, New York: Plenum 1980
176. Franck, E. U.: Pure Appl. Chem. *53*, 1401 (1981)
177. Quist, A. S. and Marshall, W. L.: J. Phys. Chem. *72*, 684 (1968)
178. Quist, A. S., Franck, E. U., Jolley, H. R. and Marshall, W. L.: J. Phys. Chem. *67*, 2453 (1963)
179. Quist, A. S. and Marshall, W. L.: J. Phys. Chem. *70*, 3714 (1966)
180. Quist, A. S. and Marshall, W. L.: J. Phys. Chem. *72*, 1545 (1968)
181. Quist, A. S. and Marshall, W. L.: J. Phys. Chem. *72*, 2100 (1968)
182. Quist, A. S. and Marshall, W. L.: J. Phys. Chem. *72*, 3122 (1968)
183. Dunn, L. A. and Marshall, W. L.: J. Phys. Chem. *73*, 723 (1969)
184. Franck, E. U.: Z. Phys. Chem. N.F. *8*, 92 (1956)
185. Franck, E. U.: Z. Phys. Chem. N.F. *8*, 107 (1956)
186. Ritzert, G. and Franck, E. U.: Ber. Bunsenges. Phys. Chem. *72*, 798 (1968)
187. Mangold, K. and Franck, E. U.: Ber. Bunsenges. Phys. Chem. *73*, 21 (1969)
188. Heger, K., Uematsu, M. and Franck, E. U.: Ber. Bunsenges. Phys. Chem. *84*, 758 (1980)
189. Fernandez-Prini, R.: in Ref. [55], p. 525
190. Spiro, M.: in Ref. [55], p. 615, 635
191. Barthel, J.: Angew. Chem. *80*, 253 (1968); Int. Ed. Engl. *7*, 260 (1968)
192. Janz, G. J. and Tomkins, R. P. T.: Nonaqueous Electrolytes Handbook, Vol. 1, New York: Academic Press (1972)
193. Justice, J.-C. and Justice, M.-C.: Faraday Discuss. Chem. Soc. *64*, 265 (1977)
194. Perié, J., Perié, M. and Justice, J.-C.: J. Solution Chem. *9*, 395 (1980); Justice, J.-C., Perié, J. and Perié, M.: J. Solution Chem. *9*, 583 (1980)
195. Barthel, J., Justice, J.-C. and Wachter, R.: Z. Phys. Chem. N.F. *84*, 100 (1973)
196. Fuoss, R. M.: J. Solution Chem. *7*, 771 (1978); J. Phys. Chem. *79*, 525, 1983 (1975); *82*, 2427 (1978)
197. Ebeling, W., Feistel, R., Kelbg, G. and Sändig, R.: J. Nonequil. Thermodyn. *3*, 11 (1978)
198. Turq, P., Orcil, L., Chemla, M. and Barthel, J.: Ber. Bunsenges. Phys. Chem. *85*, 535 (1981); Turq, P., Orcil, L., Chevalet, J., Mills, R. and Chemla, M.: in press
199. Hubbard, J. B. and Onsager, L.: J. Chem. Phys. *67*, 4850 (1977); Hubbard, J. B.: J. Chem. Phys. *68*, 1649 (1978)
200. Pikal, M. J.: J. Phys. Chem. *75*, 663 (1971)
201. Chen, M.-S. and Onsager, L.: J. Phys. Chem. *81*, 2017 (1977); Chen, M.-S.: J. Solution Chem. *7*, 675 (1978); *8*, 165, 509 (1979)
202. Quint, J. and Viallard, A.: J. Solution Chem. *7*, 137, 525, 533 (1978)
203. Lee, W. H. and Wheaton, R. J.: J. Chem. Soc., Faraday Trans. 2 *74*, 743, 1456 (1978); *75*, 1128 (1979)
204. Pethybridge, A. D. and Taba, S. S.: Faraday Discuss. Chem. Soc. *64*, 274 (1977); J. Chem. Soc., Faraday Trans. 1 *76*, 368 (1980); *78*, 1331 (1982)
205. Onsager, L.: Phys. Z. *27*, 388 (1926)
206. Fuoss, R. M. and Hirsch, E.: J. Amer. Chem. Soc. *82*, 1013 (1960)
207. Brown, A. M. and Fuoss, R. M.: J. Phys. Chem. *64*, 1341 (1960)
208. Coetzee, J. F. and Cunningham, G. P.: J. Amer. Chem. Soc. *87*, 2529 (1965)
209. Evans, D. F. and Gardam, P.: J. Phys. Chem. *72*, 3281 (1968)
210. Krumgalz, B.: Russ. J. Phys. Chem. (Engl. Transl.) *47*, 528 (1973)
211. Kay, R. L. and Dye, J. L.: Proc. Nat. Acad. Sci. USA *49*, 5 (1963); Vidulich, G. A., Cunningham, G. P. and Kay, R. L.: J. Solution Chem. *2*, 23 (1973)
212. Moynihan, C. T.: in Ref. [50], Vol. 1, p. 261
213. Angell, C. A. and Bressel, R. D.: J. Phys. Chem. *76*, 3244 (1972)
214. Angell, C. A.: J. Phys. Chem. *70*, 3988 (1966)
215. Adam, G. and Gibbs, J. H.: J. Chem. Phys. *43*, 139 (1965)
216. Cohen, M. H. and Turnbull, D.: J. Chem. Phys. *31*, 1164 (1959)
217. Angell, C. A., Sare, J. M. and Sare, E. J.: J. Phys. Chem. *82*, 2622 (1978)
218. Angell, C. A. and Tucker, J. C.: J. Phys. Chem. *78*, 278 (1974)
219. Bruno, P. and Della Monica, M.: Electrochim. Acta *20*, 179 (1975); J. Phys. Chem. *76*, 3034 (1972)
220. Bruno, P., Gatti, C. and Della Monica, M.: Electrochim. Acta *20*, 533 (1975)

221. Spiro, M. and King, F. in: Inman, D. and Lovering, D. G. (eds.): Ionic Liquids, p. 57, New York: Plenum 1981
222. Harrison, G.: The Dynamic Properties of Supercooled Liquids, New York: Academic Press 1976
223. Copeland, J. L.: Transport Properties of Ionic Liquids, Ch. 4, New York: Gordon and Breach 1974
224. Anderson, P. W. in: Balian, R., Maynard, R. and Toulouse, G. (eds.): Ill-Condensed Matter, Les Houches, Session XXXI, 3. Juillet — 18 Aout. 1978, Amsterdam: North Holland 1979
225. Leist, M.: Z. Phys. Chem. (Leipzig) *205*, 16 (1955)
226. Wishaw, B. F. and Stokes, R. H.: J. Amer. Chem. Soc. *76*, 2065 (1954)
227. Gores, H.-J. and Barthel, J.: J. Solution Chem. *9*, 939 (1980)
228. Ebeling, W., Geisler, D., Kraeft, D. and Sändig, R.: Wiss. Z. Univ. Rostock, Math.-Naturwiss. Reihe *23*, 903 (1974)
229. Jasinski, R.: High-Energy Batteries, New York: Plenum 1967
230. Jasinski, R. in: Tobias, C. W. (ed.): Advances in Electrochemistry and Electrochemical Engineering, Vol. 8, p. 253, New York: Wiley 1971
231. Casteel, J. F. and Amis, E. S.: J. Chem. Eng. Data *17*, 55 (1972)
232. Barthel, J., Gores, H.-J. and Schmeer, G.: Ber. Bunsenges. Phys. Chem. *83*, 911 (1979)
233. Barthel, J., Gerber, R. and Gores, H.-J.: in preparation
234. Fuoss, R. M. and Kraus, C. A.: J. Amer. Chem. Soc. *55*, 2387 (1933)
235. Peled, E., Brand, M. and Gileadi, E.: J. Electrochem. Soc. *128*, 1697 (1981)
236. Reger, A., Peled, E. and Gileadi, E.: J. Phys. Chem. *83*, 873 (1979)
237. Reger, A., Peled, E. and Gileadi, E.: J. Phys. Chem. *83*, 869 (1979)
238. Gorenbein, E. Y.: Russ. J. Phys. Chem. (Engl. Transl.) *20*, 547 (1946)
239. Rubinstein, I. and Gileadi, E.: J. Electrochem. Soc. *126*, 1368 (1979)
240. Barthel, J. and Wittmann, H.-J.: in preparation
241. Neilson, G. W. and Enderby, J. E.: Annual Reports C, p. 185, London: The Royal Chemical Society 1979
242. Enderby, J. E. and Neilson, G. W.: Adv. Phys. *29*, 323 (1980)
243. Ohtaki, H. and Johansson, G.: Pure Appl. Chem. *53*, 1357 (1981)
244. Stern, E. A.: Contemp. Phys. *19*, 289 (1978)
245. Sandstrom, D. R. and Lytle, F. W.: Annu. Rev. Phys. Chem. *30*, 215 (1979)
246. Eisenberger, P. and Kincaid, B. M.: Chem. Phys. Lett. *36*, 134 (1975)
247. Fontaine, A., Lagarde, P., Raoux, D., Fontana, M. P., Maisano, G., Migliardo, P. and Wanderlingh, F.: Phys. Rev. Lett. *41*, 504 (1978)
248. Sandstrom, D. R.: J. Chem. Phys. *71*, 2381 (1979)
249. Sandstrom, D. R., Dodgen, H. W. and Lytle, F. W.: J. Chem. Phys. *67*, 473 (1977)
250. Barthel, J. (ed.): 7th International Conference on Non-Aqueous Solutions, Vol. 1, Vol. 2, Regensburg 1980
251. Goulon, J. and Chabanel, M.: in Ref. [250], Vol. I, p. C1
252. Kosower, E. M.: J. Amer. Chem. Soc. *80*, 3261 (1958)
253. Libuś, W.: Pure Appl. Chem. *53*, 1345 (1981)
254. Symons, M. C. R. in: Kevan, L. and Webster, B. C. (eds.): Electron-Solvent and Anion-Solvent Interactions, p. 311, Amsterdam: Elsevier 1976
255. Hogen-Esch, T. E. and Smid, J.: J. Amer. Chem. Soc. *87*, 669 (1965)
256. Hogen-Esch, T. E. and Smid, J.: J. Amer. Chem. Soc. *88*, 307, 318 (1966)
257. Chang, C. J., Kiesel, R. F. and Hogen-Esch, T. E.: J. Amer. Chem. Soc. *97*, 2805 (1975)
258. Hogen-Esch, T. E. in: Gold, V. and Bethell, D. (eds.): Advances in Physical Organic Chemistry, Vol. 15, p. 153, New York: Academic Press 1977
259. Dubois, J.-E., Goetz, E. and Bienvenüe, A.: Spectrochim. Acta *20*, 1815 (1964); Dubois, J.-E. and Bienvenüe, A.: J. Chim. Phys. Phys.-Chim. Biol. *65*, 1259 (1968)
260. Kamlet, M. J., Abboud, J. L. M. and Taft, R. W. in: Taft, R. W. (ed.): Progress in Physical Organic Chemistry, Vol. 13, p. 485, New York: Wiley 1981
261. Ahmed, W.: J. Chem. Educ. *56*, 795 (1979)
262. Irish, D. E. and Brooker, M. H. in: Clark, R. J. H. and Hester, R. E. (eds.): Advances in Infrared and Raman Spectroscopy, Vol. 2, p. 212, London: Heyden 1976

263. Bulmer, J. T. and Shurvell, H. F.: J. Phys. Chem. *77*, 256, 2085 (1973); Bulmer, J. T. and Shurvell, H. F.: Can. J. Chem. *53*, 1251 (1975)
264. Bulmer, J. T., Irish, D. E., Grossman, F. W., Herriot, G., Tseng, M. and Weerheim, A. J.: Appl. Spectrosc. *29*, 506 (1975)
265. Gans, P.: Computers and Chemistry *1*, 291 (1977)
266. Gans, P. and Gill, J. B.: Appl. Spectrosc. *31*, 451 (1977)
267. Gans, P.: Proc. Anal. Div. Chem. Soc. *17*, 133 (1980)
268. Edgell, W. F., Schmidlin, E. and Balk, M. W.: Appl. Spectrosc. *34*, 420 (1980)
269. Kauppinen, J. K., Moffatt, D. J., Mantsch, H. H. and Cameron, D. G.: Appl. Spectrosc. *35*, 271 (1981)
270. Freeman, J. J., Heaviside, J., Hendra, P. J., Prior, J. and Reid, E. S.: Appl. Spectrosc. *35*, 196 (1981)
271. Martin, A. E. in: Durig, J. R. (ed.): Vibrational Spectra and Structure, Vol. 8: Infrared Interferometric Spectrometers, Amsterdam: Elsevier 1980
272. Coetzee, J. F. and Sharpe, W. R.: J. Solution Chem. *1*, 77 (1972)
273. Perelygin, I. S. and Klimchuk, M. A.: Russ. J. Phys. Chem. (Engl. Transl.) *47*, 1138 (1973)
274. Perelygin, I. S. and Klimchuk, M. A.: Russ. J. Phys. Chem. (Engl. Transl.) *47*, 1402 (1973)
275. Perelygin, I. S. and Klimchuk, M. A.: Russ. J. Phys. Chem. (Engl. Transl.) *48*, 363 (1974)
276. Perelygin, I. S., Klimchuk, M. A. and Beloborodowa, N. N.: Russ. J. Phys. Chem. (Engl. Transl.) *54*, 1703 (1980)
277. Chang, T. G. and Irish, D. E.: J. Solution Chem. *3*, 161 (1974)
278. Sze, Y.-K. and Irish, D. E.: J. Solution Chem. *8*, 395 (1979)
279. Janz, G. J. and Müller, M. A.: J. Solution Chem. *4*, 285 (1975)
280. Kabisch, G.: Abh. Sächs. Akad. Wiss. Leipzig, Math.-Naturwiss. Klasse, *54*, 71 (1979)
281. Tsatsas, A. T., Stearns, R. W. and Risen, Jr., W. M.: J. Amer. Chem. Soc. *94*, 5247 (1972)
282. Rannou, J. and Chabanel, M.: J. Chim. Phys. Phys.-Chim. Biol. *77*, 201 (1980)
283. Edgell, W. F. and Harris, D.: J. Solution Chem. *9*, 649 (1980)
284. Roberts, J. H., Lemley, A. T. and Lagowski, J. J.: Spectrosc. Lett. *5*, 271 (1972)
285. Lemley, A. T., Roberts, J. H., Plowman, K. R. and Lagowski, J. J.: J. Phys. Chem. *77*, 2185 (1973)
286. Plowman, K. R. and Lagowski, J. J.: J. Phys. Chem. *78*, 143 (1974)
287. Lemley, A. T. and Lagowski, J. J.: J. Phys. Chem. *78*, 708 (1974)
288. Gill, J. B.: Pure Appl. Chem. *53*, 1365 (1981)
289. Gans, P., Gill, J. B., Griffin, M. and Cahill, P. C.: J. Chem. Soc., Dalton Trans. *1981*, 968
290. Symons, M. C. R.: Pure Appl. Chem. *51*, 1671 (1979)
291. Lees, A. J., Stroughan, B. P. and Gardiner, D. J.: J. Mol. Struct. *71*, 61 (1981)
292. Bukowska, J. and Miaskiewicz, K.: J. Mol. Struct. *74*, 1 (1981)
293. Barthel, J., Behret, H. and Schmithals, F.: Ber. Bunsenges. Phys. Chem. *75*, 305 (1971)
294. Badiali, J.-P., Cachet, H. and Lestrade, J.-C.: Pure Appl. Chem. *53*, 1383 (1981)
295. Chabanel, M., Ménard, C. and Guihéneuf, G.: C.R. Acad. Sci. (Paris) *272C*, 253 (1971)
296. Paoli, D., Luçon, M. and Chabanel, M.: Spectrochim. Acta *35A*, 593 (1979)
297. Edgell, W. F., Hegde, S. and Barbetta, A.: J. Amer. Chem. Soc. *100*, 1406 (1978)
298. Edgell, W. F. and Chanjamsri, S.: J. Amer. Chem. Soc. *102*, 147 (1980)
299. Barker, C. and Yarwood, J.: J. Chem. Soc., Faraday Trans. 2 *71*, 1322 (1975)
300. Riddell, J. D., Lockwood, D. J. and Irish, D. E.: Can. J. Chem. *50*, 2951 (1972)
301. Pease, L. G., Baron, D., Easwaran, K. R. K. and Blout, E. R. in: Ananchenko, S. N. (ed.) Front. Bioorg. Chem. Mol. Biol. Proc. Int. Symp. 1978, p. 81, Oxford: Pergamon 1980
302. Hertz, H. G.: in Ref. [46], p. 479
303. von Goldammer, E. in: Bockris, J. O'M. and Conway, B. E. (eds.): Modern Aspects of Electrochemistry, Vol. 10, p. 1, New York: Plenum 1975
304. Popov, A. I.: Pure Appl. Chem. *51*, 101 (1979)
305. Harris, R. K. and Mann, B. E. (eds.): NMR and the Periodic Table, New York: Academic Press 1978
306. Briggs, R. W., Metz, K. R. and Hinton, J. F.: J. Solution Chem. *8*, 479 (1979)
307. Metz, K. R. and Hinton, J. F.: J. Solution Chem. *10*, 21 (1981)
308. Symons, M. C. R. in: Kevan, L. and Webster, B. C. (eds.): Electron-Solvent and Anion-Solvent Interactions, p. 327, Amsterdam: Elsevier 1976

309. Greenberg, M. S., Bodner, R. L. and Popov, A. I.: J. Phys. Chem. *77*, 2449 (1973)
310. Cahen, Y. M., Handy, P. R., Roach, E. T. and Popov, A. I.: J. Phys. Chem. *79*, 80 (1975)
311. Dewitte, W. J., Schoening, R. C. and Popov, A. I.: Inorg. Nucl. Chem. Lett. *12*, 251 (1976)
312. Dewitte, W. J., Liu, L., Mei, E., Dye, J. L. and Popov, A. I.: J. Solution Chem. *6*, 337 (1977)
313. Shih, J. S. and Popov, A. I.: Inorg. Nucl. Chem. Lett. *13*, 105 (1977)
314. Heubel, P.-H. and Popov, A. I.: J. Solution Chem. *8*, 283 (1979)
315. Rahimi, A. K. and Popov, A. I.: J. Magn. Res. *36*, 351 (1979)
316. Farmer, R. M. and Popov, A. I.: Inorg. Nucl. Chem. Lett. *17*, 51 (1981)
317. Buckson, R. L. and Smith, S. G.: J. Phys. Chem. *68*, 1875 (1964)
318. Cahen, Y. M., Dye, J. L. and Popov, A. I.: J. Phys. Chem. *79*, 1289 (1975)
319. Melendres, C. A. and Hertz, H. G.: J. Chem. Phys. *61*, 4156 (1974)
320. Geiger, A. and Hertz, H. G.: Adv. Mol. Relaxation Processes *9*, 293 (1977)
321. Weingärtner, H. and Hertz, H. G.: Ber. Bunsenges. Phys. Chem. *81*, 1204 (1977)
322. Singer, J. R.: J. Phys. E: Sci. Instr. *11*, 281 (1978)
323. Holz, M. and Müller, C.: Ber. Bunsenges. Phys. Chem. *86*, 141 (1982)
324. Sharp, J. H. and Symons, M. C. R. in: Szwarc, M. (ed.): Ions and Ion Pairs in Organic Reactions, Vol. 1, p. 177, New York: Wiley 1972
325. Hill, N. E., Vaughan, W. E., Price, A. H. and Davies, M.: Dielectric Properties and Molecular Behaviour, London: Van Nostrand Reinhold 1969
326. Barthel, J. and Feuerlein, F.: in preparation
327. Lestrade, J.-C., Badiali, J.-P. and Cachet, H. in: Davies, M. (ed.): Dielectric and Related Molecular Processes (Spec. Period. Rep.), Vol. 2, p. 106, London: The Chemical Society 1975
328. Davies, M. and Johansson, G.: Acta Chem. Scand. *18*, 1171 (1964)
329. Barthel, J., Schmithals, F. and Behret, H.: Z. Phys. Chem. N.F. *71*, 115 (1970); Behret, H., Schmithals, F. and Barthel, J.: Z. Phys. Chem. N.F. *96*, 73 (1975)
330. Barthel, J., Krüger, J. and Schollmeyer, E.: Z. Phys. Chem. N.F. *104*, 59 (1977)
331. Barthel, J. and Kaukal, B. in: Ref. [250], Vol. II, p. 36; Kaukal, B.: Dissertation, Regensburg 1982
332. Brot, C. in: Davies, M. (ed.): Dielectric and Related Molecular Processes (Spec. Period. Rep.), Vol. 2, p. 1, London: The Chemical Society 1975
333. Hill, N. E.: Proc. Phys. Soc. *82*, 723 (1963)
334. Edgell, W. F., Watts, A. T., Lyford, J. and Risen, Jr., W. M.: J. Amer. Chem. Soc. *88*, 181 (1966)
335. Maxey, B. W. and Popov, A. I.: J. Amer. Chem. Soc. *89*, 2230 (1967); Popov, A. I.: Pure Appl. Chem. *41*, 275 (1975)
336. Edgell, W. F. in: Szwarc, M. (ed.): Ions and Ion Pairs in Organic Reactions, Vol. 1, p. 153, New York: Wiley 1972
337. Falkenhagen, H.: in Ref. [46], p. 168
338. Ebeling, W. and Justice, J.-C. in: Ref. [64], p. 133
339. van Beek, W. M.: Dissertation, Leiden 1975
340. Irish, D. E., Chang, T. G., Tang, S.-Y. and Petrucci, S.: J. Phys. Chem. *85*, 1686 (1981)
341. Strehlow, H. and Knoche, W.: Fundamentals of Chemical Relaxations, Weinheim: Verlag Chemie 1977
342. Eigen, M. and Tamm, K.: Z. Elektrochem. *66*, 93, 107 (1962)
343. Alexander, R., Ko, E. C. F., Parker, A. J. and Broxton, T. J.: J. Amer. Chem. Soc. *90*, 5049 (1968)
344. Makosza, M. and Wawrzyniewicz, M.: Tetrahedron Lett. *1969*, 4659
345. Boden, R. M.: Synthesis *1975*, 784
346. Parker, A. J., Clare, B. W. and Smith, R. P.: Hydrometallurgy *4*, 233 (1979)
347. Livingston, R.: J. Chem. Educ. *7*, 2887 (1930)
348. Scatchard, G.: Chem. Rev. *10*, 229 (1932)
349. Perlmutter-Hayman, B. in: Jennings, K. R. and Cundall, R. B. (eds.): Progress in Reaction Kinetics, Vol. 6, p. 239, Oxford: Pergamon 1971
350. Marshall, Jr., E. K. and Acree, S. F.: J. Phys. Chem. *19*, 589 (1915)
351. Brändström, A.: Ark. Kemi *11*, 527 (1957)
352. Brändström, A.: Ark. Kemi *11*, 567 (1957)
353. Beronius, P., Nilsson, A.-M. and Wikander, G.: Acta Chem. Scand. *24*, 2826 (1970)
354. Barthel, J., Bäder, G. and Raach-Lenz, M.: Z. Phys. Chem. N.F. *103*, 135 (1976)

355. Barthel, J. and Raach-Lenz, M.: Z. Phys. Chem. N.F. *103*, 149 (1976)
356. Cayzergues, J., Georgoulis, C. and Papanastasiou, G.: J. Chim. Phys. Phys.-Chim. Biol. *74*, 1103, 1112 (1977)
357. Cayzergues, J., Georgoulis, C. and Mathieu, G.: J. Chim. Phys. Phys.-Chim. Biol. *77*, 401 (1980)
358. Dubois, J.-E. and Barthel, J.: C.R. Acad. Sci. (Paris) *245*, 1531 (1957)
359. Parker, A. J.: Chem. Rev. *69*, 1 (1969)
360. Johnson, S. L. in: Gold, V. (ed.): Advances in Physical Organic Chemistry, Vol. 5, p. 235, New York: Academic Press 1967
361. Abraham, M. H.: J. Chem. Soc., Perkin Trans. 2 *1972*, 1343
362. von Halban, H.: Z. Phys. Chem. (Leipzig) *67*, 129 (1909)
363. Menschutkin, N.: Z. Phys. Chem. (Leipzig) *6*, 41 (1890)
364. Menschutkin, N.: Z. Phys. Chem. (Leipzig) *34*, 157 (1900)
365. Yamamura, K. and Murahashi, S.: Tetrahedron Lett. *1977*, 4429
366. Hughes, E. D. and Whittingham, D. J.: J. Chem. Soc. *1960*, 806
367. Gleave, J. L.; Hughes, E. D. and Ingold, C. K.: J. Chem. Soc. *1935*, 236
368. Watts, D. W.: Pure Appl. Chem. *51*, 1713 (1979)
369. Bender, M. L.: Chem. Rev. *60*, 53 (1960)
370. Kirby, A. J. in: Bamford, C. H. and Tipper, C. F. W. (eds.): Comprehensive Chemical Kinetics, Vol. 10, p. 156, Amsterdam: Elsevier 1972
371. Kirby, A. J.: in Ref. [370], p. 162
372. Hughes, E. D. and Wilby, J.: J. Chem. Soc. *1960*, 4094
373. DePuy, C. H. and Bishop, C. A.: J. Amer. Chem. Soc. *82*, 2532 (1960)
374. Dubois, J.-E., Huynh, X. Q. and Viellard, H.: C.R. Acad. Sci. Paris *260*, 3057 (1965)
375. Garnier, F. and Dubois, J.-E.: Bull. Soc. Chim. Fr. *1968*, 3797
376. Barthel, J. and Schmeer, G. in: Dubois, J.-E. (ed.): Reaction Transition States, p. 211, New York: Gordon and Breach 1972
377. Laidler, K. J. and Landskroener, P. A.: Trans. Faraday Soc. *52*, 200 (1956)
378. Amis, E. S. and Hinton, J. F.: Solvent Effects on Chemical Phenomena, Vol. 1, New York: Academic Press 1973
379. Hiromi, K.: Bull. Chem. Soc. Japan *33*, 1251, 1264 (1960)
380. Reynaud, R.: Bull. Soc. Chim. Fr. *1968*, 2279
381. Ingold, C. K.: Structure and Mechanism in Organic Chemistry (2nd ed.), Ithaca: Cornell University Press 1969
382. Schmeer, G. and Barthel, J.: to be published
383. Caldin, E. F.: Pure Appl. Chem. *51*, 2067 (1979)
384. Ritchie, C. D.: Pure Appl. Chem. *51*, 153 (1979)
385. Svirbely, W. J. and Kuchta, A. D.: J. Phys. Chem. *65*, 1333 (1961)
386. Tommila, E.: Suom. Kemistil. *B37*, 117 (1964)
387. Goitein, R. and Bruice, T. C.: J. Phys. Chem. *76*, 432 (1972)
388. Balakrishnan, M.; Rao, G. V. and Venkatasubramanian, N.: J. Chem. Soc., Perkin Trans. 2 *1974*, 6
389. Simmons, E. L. in: Jennings, K. R. and Cundall, R. B. (eds.): Progress in Reaction Kinetics, Vol. 8, p. 161, Oxford: Pergamon 1977
390. Ritchie, C. D., Skinner, G. A. and Badding, V. G.: J. Amer. Chem. Soc. *89*, 2063 (1967)
391. Koppel, I. A. and Palm, V. A. in: Chapman, N. B. and Shorter, J. (eds.): Advances in Linear Free Energy Relationships, p. 203, New York: Plenum 1972
392. Wold, S.: Chemica Scripta *5*, 97 (1974)
393. Wold, S. and Sjöström, M. in: Chapman, N. B. and Shorter, J. (eds.): Correlation Analysis in Chemistry, p. 1, New York: Plenum 1978
394. Fainberg, A. H. and Winstein, S.: J. Amer. Chem. Soc. *78*, 2770 (1956)
395. Smith, S. G., Fainberg, A. H. and Winstein, S.: J. Amer. Chem. Soc. *83*, 618 (1961)
396. Lassau, C. and Jungers, J.-C.: Bull. Soc. Chim. Fr. *1968*, 2678
397. Drougard, Y. and Decroocq. D.: Bull. Soc. Chim. Fr. *1969*, 2972
398. Parker, A. J., Mayer, U., Schmid, R. and Gutmann, V.: J. Org. Chem. *43*, 1843 (1978)
399. Krygowski, T. M. and Fawcett, W. R.: J. Amer. Chem. Soc. *97*, 2143 (1975)
400. Fawcett, W. R. and Krygowski, T. M.: Aust. J. Chem. *28*, 2115 (1975)

401. Mayer, U.: Pure Appl. Chem. *51*, 1697 (1979)
402. Ehrenson, S.: J. Amer. Chem. Soc. *103*, 6036 (1981)
403. Kolling, O. W.: Anal. Chem. *54*, 260 (1982)
404. Kamlet, M. J., Carr, P. W., Taft, R. W. and Abraham, M. H.: J. Amer. Chem. Soc. *103*, 6062 (1981)
405. Dubois, J.-E.: Pure Appl. Chem. *53*, 1313 (1981)
406. Dubois, J.-E.: Entropie *27*, 1 (1969)
407. Fredenslund, A., Michelsen, M. L. and Sörensen, J. M.: in Ref. [71], p. 433
408. Rasmussen, P. and Skjold-Jørgensen, S.: in Ref. [71], p. 419
409. Eckermann, R.: in Ref. [71], p. 741
410. Scholz, A. G. R.: in Ref. [71], p. 761
411. Onken, U., Gmehling, J. and Arlt, W.: in Ref. [71], p. 781
412. Buck, E.: in Ref. [71], p. 771
413. Brändström, A. in: Gold, V. and Bethell, D. (eds.): Advances in Physical Organic Chemistry, Vol. 15, p. 267, London: Academic Press 1977
414. Weber, W. P. and Gokel, G. W.: Phase Transfer Catalysis in Organic Synthesis, Berlin: Springer 1977
415. Starks, C. M. and Liotta, C.: Phase Transfer Catalysis, Principles and Techniques, New York: Academic Press 1978
416. Dehmlow, E. V. and Dehmlow, S. S.: Phase Transfer Catalysis, Weinheim: Verlag Chemie 1980
417. Starks, C. M.: CHEMTECH *10*, 110 (1980)
418. Reuben, B. and Sjöberg, K.: CHEMTECH *11*, 315 (1981)
419. Dehmlow, E. V.: Chimia *34*, 12 (1980)
420. Lindblom, L. and Elander, M.: Pharm. Technol. *4* (10), 59 (1980)
421. Dehmlow, E. V., Slopianka, M. and Heider, J.: Tetrahedron Lett. *1977*, 2361
422. Juliá, S., Ginebrada, A., Guixer, J., Masana, J., Tomás, A. and Colonna, S.: J. Chem. Soc., Perkin Trans. 1 *1981*, 574
423. Haase, R., Sauermann, P.-F. and Drücker, K.-H.: Z. Phys. Chem. N.F. *47*, 224 (1965)
424. Haase, R. and Drücker, K.-H.: Z. Phys. Chem. N.F. *46*, 141 (1965)
425. Barthel, J. et al.: unpublished results
426. Mamantov, G. in: Murphy, D. W., Broadhead, J. and Steele, B. C. H. (eds.): Materials for Advanced Batteries, p. 111, New York: Plenum 1980
427. Carpio, R. A., King, L. A., Lindstrom, R. E., Nardi, J. C. and Hussey, C. L.: J. Electrochem. Soc. *126*, 1644 (1979)
428. Johnson, W. B. and Worrell, W. L.: Solid State Ionics *5*, 367 (1981)
429. Rea, J. R., Kelsey, G. S., Kuo, H. C. and Kallianidis, M.: Solid State Ionics *3/4*, 267 (1981)
430. Liang, C. C. and Bro, P.: J. Electrochem. Soc. *116*, 1322 (1969)
431. Owens, B. B.: Solid State Ionics *3/4*, 273 (1981)
432. von Alpen, U. and Bell, M. F.: Solid State Ionics *3/4*, 259 (1981)
433. Stakem, D.: Electr. Prod. Mag., Nov. (1979), p. 75
434. Graham, R. W.: Rechargeable Batteries, Advances since 1977, Chem. Techn. Rev. No. 160, Energy Techn. Rev. No. 55, Park Ridge N. J.: Noyes Data Corp. (1980)
435. Newman, G. H. and Klemann, L. P.: J. Electrochem. Soc. *127*, 2097 (1980)
436. Klemann, L. P. and Newmann, G. H.: J. Electrochem. Soc. *129*, 230 (1982)
437. Saito, T., Ikeda, H., Matsuda, Y. and Tamura, H.: J. Appl. Electrochem. *6*, 85 (1976)
438. Jasinski, R.: Electroanal. Chem. *26*, 189 (1970)
439. Eckert, J. and Gálovà, M.: Z. Chemie *20*, 194 (1980)
440. Matsuda, Y., Ōuchi, Y. and Tamura, H.: J. Appl. Electrochem. *4*, 53 (1974)
441. Takahashi, S. and Koura, N.: J. Electrochem. Soc. *128*, 710 (1981)
442. Staniewicz, R. J.: J. Electrochem. Soc. *127*, 782 (1980)
443. Scrosati, B.: Electrochim. Acta *26*, 1559 (1981)
444. Besenhard, J. O. and Heydecke, J. in: von Stumm, F. (ed.): Elektrochemie und Elektronik, p. 189, DECHEMA Monographien, Vol. 90, Weinheim: Verlag Chemie 1980
445. Murphy, D. W., Broadhead, J. and Steele, B. C. H. (eds.): Materials for Advanced Batteries, New York: Plenum 1980
446. Vissers, D. R.: in Ref. [445], p. 47

447. Huggins, R. A.: in Ref. [445], p. 91
448. Roberts, R. in: Ouellette, R. P.; Ellerbusch, F. and Cheremisinoff, P. N. (eds.): Electrotechnology, Vol. 2, p. 335, Ann Arbor: Ann Arbor Science 1978
449. Steunenberg, R. K. in: Vashishta, P., Mundy, J. N. and Shenoy, G. K. (eds.): Fast Ion Transport in Solids, p. 23, New York: North Holland 1979
450. Dell, R. M. and Bones, R. J. in: Vashishta, P., Mundy, J. N. and Shenoy, G. K. (eds.): Fast Ion Transport in Solids, p. 29, New York: North Holland 1979
451. Asher, J. A. in: Vashishta, P., Mundy, J. N. and Shenoy, G. K. (eds.): Fast Ion Transport in Solids, p. 39, New York: North Holland 1979
452. Liang, C. C. and Holmes, C. F.: in Ref. [465], p. 27
453. Greatbatch, W.: in Ref. [465], p. 3
454. Kennedy, J. H. in: Bockris, J. O'M., Conway, B. E., Yeager, E. and White, R. E. (eds.): Comprehensive Treatise of Electrochemistry, Vol. 3, p. 279, New York: Plenum 1981
455. Owens, B. B. and Skarstad, P. M. in: Vashishta, P., Mundy, J. N. and Shenoy, G. K. (eds.): Fast Ion Transport in Solids, p. 61, New York: North Holland 1979
456. Armand, M. B., Chabagno, J. M. and Duclot, M. J. in: Vashishta, P., Mundy, J. N. and Shenoy, G. K. (eds.): Fast Ion Transport in Solids, p. 131, New York: North Holland 1979
457. Kronenberg, M. L. and Blomgren, G. E. in: Bockris, J. O'M, Conway, B. E., Yeager, E. and White, R. E. (eds.): Comprehensive Treatise of Electrochemistry, Vol. 3, p. 247, New York: Plenum 1981
458. Sprengel, D.: Symposium Lithium Batterien, Kelkheim/Taunus 1980
459. Fiordiponti, P., Pistoia, G. and Temperoni, C.: J. Electrochem. Soc. *125*, 14 (1978)
460. Barthel, J., Gores, H.J. and Utz, M.: in preparation
461. Olmstead, W. N.: J. Electrochem. Soc. *127*, 344C (1980)
462. Venkatasetty, H. V. (ed.): Proceedings of the Symposium on Lithium Batteries, Pennington N.J.: The Electrochem. Soc. 1981
463. Olmstead, W. N.: in Ref. [462], p. 144
464. Angres, I. A.: J. Electrochem. Soc. *127*, 344C (1980); Angres, I. A. and James, S. D. in Ref. [465], p. 332
465. Owens, B. B. and Margalit, N. (eds.): Proceedings of the Symposia on Power Sources for Biomedical Implantable Applications and Ambient Temperature Lithium Batteries, Princeton: The Electrochemical Society 1980
466. Margalit, N. and Canning, H. J.: J. Electrochem. Soc. *127*, 344C (1980)
467. Venkatasetty, H. V., Saathoff, D. J. and Patel, B. K.: J. Electrochem. Soc. *127*, 344C (1980); Ref. [462], p. 155
468. Kronenberg, M. L.: U.S. Pat. 3,871,916 (1975); 3,951,685 (1976); Ger. Pat. 2,516,035 (1977)
469. Lehmann, G. and Gabano, J. P.: Ger. Pat. 2,154,092 (1972)
470. Whitney, T. A. and Klemann, L. P.: Ger. Offen. 2,828,628 (1979); U.S. Pat. 4,117,213 (1978); U.S. Pat. 4,104,450 (1978)
471. Klemann, L. P., Newman, G. H. and Stogryn, E. L.: U.S. Pat. 4,139,681 (1979)
472. Klemann, L. P. and Newman, G. H.: J. Electrochem. Soc. *128*, 13 (1981)
473. Newman, G. H., Francis, R. W., Gaines, L. H. and Rao, B. M. L.: J. Electrochem. Soc. *127*, 2025 (1980)
474. Matsuda, Y., Nakashima, H., Morita, M. and Takasu, Y.: J. Electrochem. Soc. *128*, 2552 (1981)
475. Matsuda, Y. and Satake, H.: J. Electrochem. Soc. *127*, 877 (1980)
476. Yeager, E. B., Schumm Jr., B., Blomgren, B., Blankenship, D. R., Leger, V. and Akridge, J. (eds.): Lithium Nonaqueous Battery Electrochemistry, Pennington: The Electrochemical Society 1980
477. Werblan, L. and Lesiński, J.: Pol. J. Chem. *53*, 2571 (1979)
478. Kedrinskii, I. A., Morozov, S. V., Sukhova, G. I. and Sokolov, L. A.: Sov. Electrochem. *12*, 1094 (1977)
479. Selim, R. and Bro, P.: J. Electrochem. Soc. *121*, 1457 (1974)
480. Dey, A. N. and Sullivan, B. P.: J. Electrochem. Soc. *117*, 222 (1970)
481. Eichinger, G.: J. Electroanal. Chem. Interfacial Electrochem. *74*, 183 (1976)
482. Dousek, F. P., Jansta, J. and Říha, J.: J. Electroanal. Chem. Interfacial Electrochem. *46*, 281 (1973)
483. Dey, A. N.: Thin Solid Films *43*, 131 (1977)

484. Epelboin, I., Froment, M., Garreau, M., Thevenin, J. and Warin, D.: J. Electrochem. Soc. *127*, 2100 (1980)
485. Seiger, H. N., Lyall, A. E. and Chair, R. C. in: Collins, D. H. (ed.): Power Sources, Vol. 2, Proc. 6th Intern. Symp. Brighton, Sept. 1968, Oxford: Pergamon 1970
486. Dey, A. N. and Holmes, R. W.: J. Electrochem. Soc. *127*, 1877 (1980)
487. Dousek, F. P. and Jansta, J.: J. Electroanal. Chem. Interfacial Electrochem. *74*, 195 (1976)
488. Caiola, A., Fadou, J. Y. and Sohm, J. C.: Electrochim. Acta *17*, 1401 (1972)
489. Caiola, A., Guy, H. and Sohm, J. C.: Electrochim. Acta *15*, 555 (1970)
490. Dey, A. N.: J. Electrochem. Soc. *127*, 1886 (1980)
491. Peled, E.: J. Electrochem. Soc. *126*, 2047 (1979)
492. Brummer, S. B., Koch, V. R. and Rauh, R. D.: in Ref. [445], p. 123
493. Clincspoor, C. C.: Dissertation, Bonn 1970
494. Lautié, R.: Bull. Soc. Chim. Fr. *118*, 508 (1947)
495. Kruger, F. J.: Symposium Lithium Batterien, Kelkheim (Taunus), 1980
496. Rasmussen, R. M.: Microtechniques, Symp. Intern. 23.–24. 4. 1981, Mulhouse (France)
497. Kruger, F. J.: Microtechniques, Symp. Intern. 23.–24. 4. 1981, Mulhouse (France)
498. Ideka, H., Narukawa, S., Inokudi, H. and Nakaido, S.: Microtechniques, Symp. Intern. 23.–24. 4. 1981, Mulhouse (France)
499. Datenblatt VARTALITH, VARTA-Batterie, AG, D-3000 Hannover 21, Postfach 210540 (1980)
500. Linden, D. and McDonald, B.: J. Power Sources *5*, 35 (1980)
501. Schlaikjer, C. R., Goebel, F. and Marincic, N.: J. Electrochem. Soc. *126*, 513 (1979)
502. Dey, A. N. and Schlaikjer, C. R.: Proc. of the 26th Power Sources Conference, Atlantic City, N.J. (1974)
503. Datenblatt Silberkraft, Leichtakkumulatoren GmbH, D-4100 Duisburg, Meidericher Str. 6–8 (1981)
504. Collins, D. H. (ed.): Power Sources, Vol. 5, Proc. 9th Intern. Symp., Brighton September 1974, London: Academic Press 1975
505. Bro, P., Holmes, R. W., Marincic, N. and Taylor, H.: in Ref. [504], p. 703
506. Gardner, C. L., Fouchard, D. T. and Fawcett, W. R.: J. Electrochem. Soc. *128*, 2337 (1981)
507. Gardner, C. L., Fouchard, D. T. and Fawcett, W. R.: J. Electrochem. Soc. *128*, 2345 (1981)
508. Dey, A. N. and Holmes, R. W.: J. Electrochem. Soc. *126*, 1637 (1979)
509. Dey, A. N. and Holmes, R. W.: J. Electrochem. Soc. *127*, 775 (1980)
510. Dey, A. N.: J. Electrochem. Soc. *127*, 1000 (1980)
511. Blomgren, G. E. and Kronenberg, M. L.: Ger. Offen. 2,262,256 (1973)
512. Behl, W. K., Christopulos, J. A., Ramirez, M. and Gilman, S.: J. Electrochem. Soc. *120*, 1619 (1973)
513. Auborn, J. J., French, K. W., Lieberman, S. I., Shah, V. K. and Heller, A.: J. Electrochem. Soc. *120*, 1613 (1973)
514. Datenblatt 1102, Sonnenschein Accumulatorenfabrik, D-6470 Büdingen, Thiergarten
515. Bowden, W. L. and Dey, A. N.: J. Electrochem. Soc. *127*, 1419 (1980)
516. Bowden, W. L. and Dey, A. N.: J. Electrochem. Soc. *126*, 2035 (1979)
517. Abraham, K. M. and Mank, R. M.: J. Electrochem. Soc. *127*, 2091 (1980)
518. Venkatasetty, H. V. and Saathoff, D. J.: J. Electrochem. Soc. *128*, 773 (1981)
519. Salomon, M.: J. Electrochem. Soc. *128*, 233 (1981)
520. Klinedinst, K. A. and Domeniconi, M. J.: J. Electrochem. Soc. *127*, 539 (1980)
521. Dey, A. N. and Miller, J.: J. Electrochem. Soc. *126*, 1445 (1979)
522. Ohzuku, T., Wakamatsu, H., Takehara, Z. and Yoshizawa, S.: Electrochim. Acta *24*, 723 (1979)
523. Gilman, S. and Wade Jr., W.: J. Electrochem. Soc. *127*, 1427 (1980)
524. Rauh, R. D., Abraham, K. M., Pearson, G. F., Surprenant, J. K. and Brummer, S. B.: J. Electrochem. Soc. *126*, 523 (1979)
525. Rauh, R. D., Shuker, F. S., Marston, J. M. and Brummer, S. B.: J. Inorg. Nucl. Chem. *39*, 1761 (1977)
526. Whittingham, M. S.: Science *192*, 1126 (1976)

527. Koch, V. R. and Young, J. H.: Science *204*, 499 (1979)
528. Holleck, G. L., Abraham, K. M. and Brummer, S. B. in: Ref. [465], p. 384
529. Journé, J. and Tobias, C. W.: J. Appl. Electrochem. *5*, 279 (1975)
530. Rauh, R. D. and Brummer, S. B.: Electrochim. Acta *22*, 75 (1977)
531. Dampier, F. W. and Brummer, S. B.: Electrochim. Acta *22*, 1339 (1977)
532. Koch, V. R. and Brummer, S. B.: Electrochim. Acta *23*, 55 (1978)
533. Rauh, R. D. and Brummer, S. B.: Electrochim. Acta *22*, 85 (1977)
534. Gunther, R. G.: U.S. Pat. 3,928,070 (1975); Ger. Offen. 2,502,497 (1975)
535. Broadhead, J. and Trumbore, F. A.: in Ref. [504], p. 661
536. Soffer, A.: U.S. Pat. 4,132,837 (1979)
537. Koch, V. R. and Young, J. H.: J. Electrochem. Soc. *125*, 1371 (1978)
538. Koch, V. R.: J. Electrochem. Soc. *126*, 181 (1979)
539. Koch, V. R.: U.S. Pat. 4,118,550 (1978)
540. Goldman, J. L., Mank, R. M., Young, J. H. and Koch, V. R.: J. Electrochem. Soc. *127*, 1461 (1980)
541. Carjaval, C., Tölle, K. J., Smid, J. and Szwarc, M.: J. Amer. Chem. Soc. *87*, 5545 (1965)
542. Nichols, D., Sutphen, C. and Szwarc, M.: J. Phys. Chem. *72*, 1021 (1968)
543. Abraham, K. M., Goldman, J. L. and Dempsey, M. D.: J. Electrochem. Soc. *128*, 2493 (1981)
544. Koch, V. R., Goldman, J. L., Mattos, C. J. and Mulvaney, M.: J. Electrochem. Soc. *129*, 1 (1982)
545. Koch, V. R.: U.S. Pat. 4,252,876 (1981)
546. Whittingham, M. S.: Progr. Solid State Chem. *12*, 41 (1978)
547. Armand, M. B.: in Ref. [445], p. 145
548. Holleck, G. L. and Driscoll, J. R.: Electrochim. Acta *22*, 647 (1977)
549. Thompson, A. H. and Symon, C. R.: Solid State Ionics, *3/4*, 175 (1981)
550. Rao, B. M. L. and Klemann, L. P.: J. Electrochem. Soc. *127*, 761 (1980)
551. Murphy, D. W., Christian, P. A., DiSalvo, F. J. and Carides, J. N.: J. Electrochem. Soc. *126*, 497 (1979)
552. Spurdens, P. C., Drennan, J., Owen, J. R., Steele, B. C. H., Gonzales-Calbet, J. M. and Jefferson, D. A.: Solid State Ionics *5*, 335 (1981)
553. Dickens, P. G. and Reynolds, G. J.: Solid State Ionics *5*, 331 (1981)
554. Jacobson, A. J.: Solid State Ionics *5*, 65 (1981)
555. Christian, P. A., Carides, J. N., DiSalvo, F. J. and Waszczak, J. V.: J. Electrochem. Soc. *127*, 2315 (1980)
556. Mizushima, K., Jones, P. C., Wiseman, P. J. and Goodenough, G. J.: Solid State Ionics *3/4*, 171 (1981)
557. Whittingham, M. S. and Newman, G. H.: J. Electrochem. Soc. *128*, 706 (1981)
558. Moshtev, R. V., Manev, V., Nassalevska, A., Pistoia, G. and Icovi, M.: J. Electrochem. Soc. *128*, 1399 (1981)
559. Cava, R. J., Santoro, A., Murphy, D. W., Zahurak, S. and Roth, R. S.: Solid State Ionics *5*, 323 (1981)
560. Eisenberg, M.: J. Electrochem. Soc. *127*, 2382 (1980)
561. Eisenberg, M.: Electrochim. Acta *26*, 955 (1981)
562. Eisenberg, M.: U.S. Pat. 4,136,223 (1979)
563. Jacobson, A. J. and Rich, S. M.: J. Electrochem. Soc. *127*, 779 (1980)
564. Jacobson, A. J., Chianelli, R. R. and Whittingham, M. S.: J. Electrochem. Soc. *126*, 2277 (1979)
565. Jacobson, A. J., Chianelli, R. R., Rich, S. M. and Whittingham, M. S.: Mat. Res. Bull. *14*, 1437 (1979)
566. Jacobson, A. J., Whittingham, M. S. and Rich, S. M.: J. Electrochem. Soc. *126*, 887 (1979)
567. Lazzari, M. and Scrosati, B.: J. Electrochem. Soc. *127*, 773 (1980)
568. MacDiarmid, A. G. and Heeger, A. J.: Synth. Met. *1*, 101 (1979/1980)
569. MacInnes, Jr., D., Druy, M. A., Nigrey, P. J., Nairns, D. P., MacDiarmid, A. G. and Heeger, A. J.: J. Chem. Soc. Chem. Commun. *1981*, 317
570. Nigrey, P. J., MacDiarmid, A. G. and Heeger, A. J.: J. Chem. Soc. Chem. Commun. *1979*, 594

571. Nigrey, P. J., MacInnes, Jr., D., Nairns, D. P., MacDiarmid, A. G. and Heeger, A. J.: J. Electrochem. Soc. *128*, 1651 (1981)
572. Thompson, J. (ed.): Power Sources, Vol. 7, Proc. 11th Intern. Symp., Brighton Sept. 1978, London: Academic Press 1979
573. Collins, D. H. (ed.): Power Sources, Vol. 6, Proc. 10th Intern. Symp., Brighton September 1976, London: Academic Press 1977
574. Besenhard, J. O. and Eichinger, G.: J. Electroanal. Chem. Interfacial Electrochem. *68*, 1 (1976)
575. Eichinger, G. and Besenhard, J. O.: J. Electroanal. Chem. Interfacial. Electrochem. *72*, 1 (1976)
576. Beghi, G. (ed.): Energy Storage and Transportation, Dordrecht: Reidel 1981
577. Popovych, O. and Tomkins, R. P. T.: Nonaqueous Solution Chemistry, New York: Wiley 1981
578. Graham, R. W.: Primary Batteries, Recent Advances, Chem. Techn. Rev. No. 105, Energy Techn. Rev. No. 25, Park Ridge, N. J.: Noyes Data Corp. 1978
579. Graham, R. W.: Secondary Batteries, Recent Advances, Chem. Techn. Rev. No. 106, Energy Techn. Rev. No. 26, Park Ridge, N. J.: Noyes Data Corp. 1978
580. Gottesfeld, S. and McIntyre, J. D. E.: J. Electrochem. Soc. *126*, 742 (1979)
581. Novotny, V. and Hopper, M. A.: J. Electrochem. Soc. *126*, 2211 (1979)
582. Camlibel, I., Singh, S., Stocker, H. J., VanUitert, L. G. and Zydzik, G. J.: Appl. Phys. Lett. *33*, 793 (1978)
583. Beni, G.: Solid State Ionics *3/4*, 157 (1981)
584. Deb, S. K.: Phil. Mag. *27*, 801 (1973)
585. Faughnan, B. W., Crandall, R. S. and Heyman, P. M.: RCA Rev. *36*, 177 (1975)
586. Chang, I. F., Gilbert, B. L. and Sun, T. I.: J. Electrochem. Soc. *122*, 955 (1975)
587. Hersh, H. N., Kramer, W. E. and McGee, J. H.: Appl. Phys. Lett. *27*, 646 (1975)
588. Faughnan, B. W., Crandall, R. S. and Lampert, M. A.: Appl. Phys. Lett. *27*, 275 (1975)
589. Crandall, R. S. and Faughnan, B. W.: Appl. Phys. Lett. *28*, 95 (1976)
590. Faughnan, B. W. and Crandall, R. S.: Appl. Phys. Lett. *31*, 834 (1977)
591. Mohapatra, S. K.: J. Electrochem. Soc. *125*, 284 (1978)
592. Reichmann, B. and Bard, A. J.: J. Electrochem. Soc. *126*, 583 (1979)
593. Randin, J. P.: J. Electron. Mat. *7*, 47 (1978)
594. Reichmann, B. and Bard, A. J.: J. Electrochem. Soc. *126*, 2133 (1979)
595. Ho, C., Raistrick, I. D. and Huggins, R. A.: J. Electrochem. Soc. *127*, 343 (1980)
596. Dautremont-Smith, W. C., Green, M. and Kang, K. S.: Electrochim. Acta *22*, 751 (1977)
597. Zeller, H. R. and Beyeler, H. U.: Appl. Phys. *13*, 231 (1977)
598. Dickens, P. G., Moore, J. H. and Neild, D. J.: Solid State Chem. *7*, 241 (1973)
599. Faughnan, B. W. and Crandall, R. S. in: Pankove, J. I. (ed.): Topics in Applied Physics, Vol. 40, Display Devices, Berlin: Springer 1980
600. Freller, H. and Mund, K. in: von Stumm, F. (ed.): Elektrochemie und Elektronik, DECHEMA-Monographien, Vol. 90, p. 107, Weinheim: Verlag Chemie 1981
601. McIntyre, J. D. E., Basu, S., Peck, Jr., W. F., Brown, W. L. and Augustyniak, W. M.: Solid State Ionics *5*, 359 (1981)
602. Burke, L. D. and O'Sullivan, E. J. M.: J. Electroanal. Chem. Interfacial Electrochem. *111*, 383 (1980)
603. Gottesfeld, S.: J. Electrochem. Soc. *127*, 272 (1980)
604. Reichmann, B. and Bard, A. J.: J. Electrochem. Soc. *127*, 241 (1980)
605. Gottesfeld, S., McIntyre, J. D. E., Beni, G. and Shay, J. L.: Appl. Phys. Lett. *33*, 208 (1978)
606. Beni, G. and Shay, J. L.: Appl. Phys. Lett. *33*, 567 (1978)
607. Shay, J. L., Beni, G. and Schiavone, L. M.: Appl. Phys. Lett. *33*, 942 (1978)
608. Schiavone, L. M., Dautremont-Smith, W. C., Beni, G. and Shay, J. L.: Appl. Phys. Lett. *35*, 823 (1979)
609. Corker, G. A., Grant, B. and Clecak, N. J.: J. Electrochem. Soc. *126*, 1339 (1979)
610. Walton, D., Ely, B. and Elliott, G.: J. Electrochem. Soc. *128*, 2479 (1981)
611. Barna, G. G.: J. Electrochem. Soc. *127*, 1317 (1980)

612. Matsumoto, S., Kato, H. and Tomura, S.: Jpn. Kokai Tokkyo Kohu 76,160,573; C.A. 93, P 58277 g (1980)
613. Miyamura, M., Tomura, S., Imai, A. and Inomata, S.: Solid State Ionics *3/4*, 149 (1981)
614. Green, M. and Kang, K.: Solid State Ionics *3/4*, 141 (1981)
615. Kmetz, A. R. in: Kmetz, A. R. and von Willisen, F. K. (eds.): Nonemissive Electrooptic Displays, p. 261, New York: Plenum 1976
616. Beni, G.: J. Electrochem. Soc. *127*, 467C (1980)
617. Bogenschütz, A. F. and Krusemark, W.: Elektrochemische Bauelemente, Weinheim: Verlag Chemie 1976
618. Dalisa, A. L.: Proc. Soc. Inf. Displ. *18*, 43 (1977)
619. Dalisa, A. L. in: Pankove, J. I. (ed.): Topics in Applied Physics, Vol. 40, Display Devices, p. 213, Berlin: Springer 1980
620. Novotny, V. and Hopper, M. A.: J. Electrochem. Soc. *126*, 925 (1979)
621. Chang, I. F. in: Kmetz, A. R. and von Willisen, F. K. (eds.): Nonemissive Electrooptic Displays, p. 155, New York: Plenum 1976
622. Pankove, J. I. in: Pankove, J. I. (ed.): Topics in Applied Physics, Vol. 40, Display Devices, p. 1, Berlin: Springer 1980
623. Matsuhiro, K. and Masuda, Y.: SID Digest 1979, quoted after Ref. [583]
624. Raistrick, I. D., Mark, A. J. and Huggins, R. A.: Solid State Ionics *5*, 351 (1981)
625. Schiavone, L. M., Dautremont-Smith, W. C., Beni, G. and Shay, J. L.: Appl. Phys. Lett. *35*, 823 (1979)
626. McIntyre, J. D. E., Peck, W. F. and Nakahara, S.: J. Electrochem. Soc. *127*, 1264 (1980)
627. Dautremont-Smith, W. C., Beni, G., Schiavone, L. M. and Shay, J. L. in: Vashishta, P., Mundy, J. N. and Shenoy, G. K. (eds.): Fast Ion Transport in Solids, p. 99, New York: North Holland 1979
628. Morrison, S. R.: Electrochemistry at Semiconductor and Oxidized Metal Electrodes, New York: Plenum 1980
629. Johnston, Jr., W. D.: Solar Voltaic Cells, New York: Dekker 1980
630. Kazmerzki, L. in: Murr, L. E. (ed.): Solar Materials Science, p. 525, New York: Academic Press 1980
631. Photoelectrochemistry, Faraday Discuss. Chem. Soc. *70* (1981)
632. Ang, P. G. P. and Sammells, A. F.: in Ref. [631], p. 207
633. Gerischer, H.: in Ref. [631], p. 137
634. Gerischer, H.: in Seraphin, B. O. (ed.): Topics in Applied Physics, Vol. 31, Solar Energy Conversion, p. 115, Berlin: Springer 1979
635. Tributsch, H.: in Ref. [631], p. 189
636. Bockris, J. O'M.: in Ref. [631], p. 429
637. Cheremisinoff, P. N. and Regino, T. C.: Principles and Applications of Solar Energy, Ann. Arbor: Ann. Arbor Science 1978
638. Weaver, N. L., Singh, R., Rajeshwar, K., Singh, P. and Dubow, J.: Solar Cells *3*, 221 (1981)
639. Park, S. M. and Barber, M. E.: J. Electroanal. Chem. Interfacial Electrochem. *99*, 67 (1979)
640. Wilson, J. R. and Park, S.-M.: J. Electrochem. Soc. *129*, 149 (1982)
641. Gerischer, H.: J. Electroanal. Chem. Interfacial Electrochem. *82*, 133 (1977)
642. Gerischer, H.: J. Vac. Sci. Technol. *15*, 1422 (1978)
643. Bard, A. J. and Wrighton, M. S.: J. Electrochem. Soc. *124*, 1706 (1977)
644. Kohl, P. A. and Bard, A. J.: J. Electrochem. Soc. *126*, 603 (1979)
645. Langmuir, M. E., Hoenig, P. and Rauh, R. P.: J. Electrochem. Soc. *128*, 2357 (1981)
646. Kohl, P. A. and Bard, A. J.: J. Electrochem. Soc. *126*, 598 (1979)
647. Curran, J. S.: in Ref. [631], p. 259
648. Cardon, F., Gomes, W. P., VanDen Kerchove, F., Vanmeakelbergh, D. and Van Overmeire, F.: in Ref. [631], p. 153
649. Kawai, T., Tributsch, H. and Sakata, T.: Chem. Phys. Lett. *69*, 336 (1980)
650. Fan, F.-R. F., White, H. S., Wheeler, B. and Bard, A. J.: J. Electrochem. Soc. *127*, 518 (1980)
651. Kline, G., Kam, K., Canfield, D. and Parkinson, B. A.: Sol. Energy Mater. *4*, 301 (1981)
652. Ang, P. G. P. and Sammells, A. F.: J. Electrochem. Soc. *129*, 233 (1982)

653. Ginley, D. S., Biefeld, R. M., Parkinson, B. A. and Keung-Kam, K.: J. Electrochem. Soc. *129*, 145 (1982)
654. Tenne, R.: J. Electrochem. Soc. *129*, 143 (1982)
655. Tributsch, H.: Ber. Bunsenges. Phys. Chem. *81*, 361 (1977)
656. Tributsch, H.: Ber. Bunsenges. Phys. Chem. *82*, 169 (1978)
657. Bard, A. J.: J. Phys. Chem. *86*, 172 (1982)
658. Noufi, R. and Tench, D.: J. Electrochem. Soc. *127*, 188 (1980)
659. Noufi, R., Tench, D. and Warren, L. F.: J. Electrochem. Soc. *127*, 2709 (1980)
660. Noufi, R., Tench, D. and Warren, L. F.: J. Electrochem. Soc. *128*, 2363 (1981)
661. Laser, D. and Bard, A. J.: J. Phys. Chem. *80*, 459 (1976)
662. Kohl, P. A. and Bard, A. J.: J. Electrochem. Soc. *126*, 59 (1979)
663. Kohl, P. A. and Bard, A. J.: J. Amer. Chem. Soc. *99*, 7531 (1977)
664. Noufi, R., Tench, D. and Warren, L. F.: J. Electrochem. Soc. *127*, 2310 (1980)
665. Noufi, R., Tench, D. and Warren, L. F.: J. Electrochem. Soc. *128*, 2596 (1981)
666. Diaz, A. F., Kanazawa, K. K. and Gardini, G. P.: J. Chem. Soc. Chem. Commun. *1979*, 635
667. Kanazawa, K. K., Diaz, A. F., Geiss, R. H., Gill, W. D., Kwak, J. F., Logan, J. A., Rabolt, J. F. and Street, G. B.: J. Chem. Soc. Chem. Commun. *1979*, 854
668. Diaz, A. F. and Castillo, J. I.: J. Chem. Soc. Chem. Commun. *1980*, 397
669. Neville, R. C.: Solar Energy Conversion: The Solar Cell, Amsterdam: Elsevier 1978
670. Heller, A. (ed.): Semiconductor Liquid Junction Cells, Pennington N.J.: The Electrochemical Society 1977
671. Seraphin, B. O. (ed.): Topics in Applied Physics, Vol. 31, Solar Energy Conversion, Berlin: Springer 1979
672. Brenner, A. in: Tobias, C. W. (ed.): Advances in Electrochemistry and Electrochemical Engineering, Vol. 5, p. 205, New York: Wiley 1967
673. Lowenheim, F. A. in: Vossen, J. L. and Kern, W. (eds.): Thin Film Processes, p. 209, New York: Academic Press 1978
674. Takei, T.: Surf. Techn. *8*, 543 (1979)
675. Couch, D. E.: Plating *49*, 363 (1962)
676. Beach, J. G. and Faust, C. L.: J. Electrochem. Soc. *106*, 654 (1959)
677. Yoshio, M. and Ishibashi, N.: J. Appl. Electrochem. *3*, 321 (1973)
678. Roethlein, R. J.: J. Electrochem. Soc. *117*, 931 (1970)
679. Schmidt, F. J. and Hess, I. J.: Plating *53*, 229 (1966)
680. Gaugin, R. and Nury, G.: Bull. Soc. Fr. Electriciens, 7e série, *8*, 733 (1958)
681. Takei, T.: Surf. Techn. *9*, 285 (1979)
682. Elwell, D.: J. Cryst. Growth *52*, 741 (1981)
683. Isserlis, G. in: Kuhn, A. T. (ed.): Industrial Electrochemical Processes, p. 376, Amsterdam: Elsevier 1971
684. Couch, D. E. and Brenner, A.: J. Electrochem. Soc. *99*, 234 (1952)
685. Connor, J. H. and Brenner, A.: J. Electrochem. Soc. *103*, 657 (1956)
686. Beach, J. G., McGraw, L. D. and Faust, C. L.: Plating *55*, 936 (1968)
687. Ishibashi, N. and Yoshio, M.: Electrochim. Acta *17*, 1343 (1972)
688. Yoshio, M., Ishibashi, N., Waki, W. and Seiyama, T.: J. Inorg. Nucl. Chem. *34*, 2439 (1972)
689. Eckert, J. and Gálovà, M.: Electrochim. Acta *26*, 1169 (1981)
690. Gálovà, M., Lux, L. and Eckert, J.: Z. Phys. Chem. (Leipzig) *263*, 377 (1982)
691. Peled, E., Mitavski, A., Reger, A. and Gileadi, E.: J. Electroanal. Chem. Interfacial Electrochem. *75*, 677 (1977)
692. Peled, E. and Gileadi, E.: Plating *62*, 342 (1975)
693. Peled, E. and Gileadi, E.: J. Electrochem. Soc. *123*, 15 (1976)
694. Peled, E., Mitavski, A. and Gileadi, E.: Z. Phys. Chem. N.F. *98*, 111 (1975)
695. Reger, A., Peled, E. and Gileadi, E.: J. Electrochem. Soc. *123*, 638 (1976)
696. Ziegel, S., Peled, E. and Gileadi, E.: Electrochim. Acta *23*, 363 (1978)
697. Yoshio, M., Nakamura, H., Nogouchi, N. and Nagamatsu, M.: in Ref. [250], p. E 3
698. Anonymous: Chem. Eng. News, *57*, May 14, p. 8, 1979

699. Barbier, M. and Bovarnick, B. in: Ouelette, R. P., Ellerbusch, F. and Cheremisinoff, P. N. (eds.): Electrotechnology, Vol. 2, p. 259, Ann Arbor: Ann Arbor Science 1978
700. Biallozor, S. and Lisowska, A.: Electrochim. Acta *25*, 1209 (1980)
701. Reid, Jr., W. E., Bish, J. M. and Brenner, A.: J. Electrochem. Soc. *104*, 21 (1957)
702. Santos, F. and Dyment, F.: Plating *60*, 821 (1973)
703. Fischer, H. W. and Schwabe, K.: Korrosion (Dresden) *11*, 105 (1980)
704. Agrawal, A. K. and Austin, A. E.: J. Electrochem. Soc. *128*, 2292 (1981)
705. Rama Mohan, T. R. and Kroeger, F. A.: Electrochim. Acta *27*, 371 (1982)
706. Bernard, W. J. in: Hampel, C. A. (ed.): Encyclopedia of Electrochemistry, p. 140, New York: Reinhold 1964
707. Morley, A. R. and Campbell, D. S.: Radio Electron. Eng. *43*, 421 (1973)
708. Campbell, D. S.: Radio Electron. Eng. *41*, 5 (1971)
709. Vijh, A. K. in: Diggle, J. W. (ed.): Oxides and Oxide Films, Vol. 2, New York: Dekker 1973
710. Campbell, D. S. in: Chapman, B. N. and Anderson, J. C. (eds.): Science and Technology of Surface Coating, p. 87, London: Academic Press 1974
711. Lacour, H. R.: Elektronische Bauelemente, Vol. 1, Stuttgart: Berliner Union 1978
712. Wood, G. C. in: Diggle, J. W. (ed.): Oxide and Oxide Films, Vol. 2, New York: Dekker 1973
713. Westwood, W. D., Waterhouse, N. and Wilcox, P. S.: Tantalum Thin Films, London: Academic Press 1975
714. Young, L.: Can. J. Chem. *38*, 1141 (1960)
715. Hagihara, M. H. and Takimoto, S. M.: Ger. Pat. 1,614,166 (1973)
716. Höft, H.: Passive Elektronische Bauelemente, Heidelberg: Hüthig 1977
717. Taketani, Y. D. and Niwa, S.: Ger. Offen, 2,508,904 (1978)
718. Pringuet, M. G.: Ger. Pat. 2,449,282 (1975)
719. Jackson, N. F. and Sethi, R. S.: Brit. Pat. 1,445,894 (1976)
720. Mallory, P. R.: Ger. Offen. 2,646,500 (1977)
721. Anderson, D. J.: U.S. Pat. 3,812,038 (1974)
722. Zinke, O.: Widerstände, Kondensatoren, Spulen und ihre Werkstoffe, Berlin: Springer 1965
723. Burger, F. J. and Young, L.: Progress in Dielectrics *5*, 1 (1962)
724. Vermileya, D. A. in: Delahay, P. (ed.): Advances in Electrochemistry and Electrochemical Engineering, Vol. 3, p. 211, New York: Wiley 1963
725. Burger, F. J. and Wu, J. C.: J. Electrochem. Soc. *118*, 2039 (1971)
726. Yahalom, J. and Hoar, T. P.: Electrochim. Acta *15*, 877 (1970)
727. Ikonopisov, S.: Electrochim. Acta *22*, 1077 (1977)
728. Ikonopisov, S., Girginov, A. and Machkova, M.: Electrochim. Acta *22*, 1283 (1977)
729. Ikonopisov, S., Girginov, A. and Machkova, M.: Electrochim. Acta *24*, 451 (1979)
730. Tajima, S.: Electrochim. Acta *22*, 995 (1977)
731. Mund, K. and Richter, R. in: von Stumm, F. (ed.): DECHEMA-Monographien, Vol. 90, Elektrochemie und Elektronik, p. 173, Weinheim: Verlag Chemie 1981
732. Sanada, K. and Hosokawa, M.: NEC-Res. and Development *55*, 21 (1979)
733. Krumpelt, M., Weissmann, E. Y. and Alkire, R. C. (eds.): Electro-organic Synthesis Technology, AIChE Symp. Ser. No. 185, Vol. 75, AIChE National Meeting, Atlanta, USA, 26. 2.–1. 3. 1978. New York: AIChE 1979
734. Swann, Jr., S. and Alkire, R. C.: in Ref. [733], p. 61
735. Swann, Jr., S. and Alkire, R. C.: Bibliography of Electro-Organic Synthesis 1801–1975, Princeton: The Electrochemical Society 1980
736. Danly, D. in: The Kirk-Othmer Encyclopedia of Chemical Technology, Vol. 8 (3rd ed.), p. 696, New York: Wiley 1979
737. Beck, T. R., Alkire, R. C. and Weinberg, N. L.: J. Electrochem. Soc. *126*, 136 C (1979)
738. Weinberg, N. L.: in Ref. [733], p. 31
739. Fitzjohn, J. L.: in Ref. [733], p. 65
740. Bockris, J. O'M. in: Bockris, J. O'M., Rand, D. A. J. and Welch, B. J. (eds.): Trends in Electrochemistry, New York: Plenum 1977
741. Alkire, R. C. and Gould, R. M.: J. Electrochem. Soc. *127*, 605 (1980)
742. Heinze, J.: Angew. Chem. *93*, 186 (1981)

743. Rifi, M. R. and Covitz, F. H.: Introduction to Organic Electrochemistry, New York: Dekker 1974
744. Pietsch, S. J. and Langer, S. H.: in Ref. [733], p. 51
745. Weinberg, N. L. (ed.) in: Weissberger, A. (ser. ed.): Techniques of Chemistry, Vol. 5, Technique of Electro-organic Synthesis, Part I (1974), Part II (1975), Part III (1982). New York: Wiley
746. Funt, B. L. and Tanner, J. in Ref. [745], Part II, p. 559
747. Houghton, R. W. and Kuhn, A. T.: J. Appl. Electrochem. *4*, 173 (1974)
748. Newman, J. S. and Tiedemann, W. in: Gerischer H. and Tobias, C. W. (eds.): Advances in Electrochemical Engineering, Vol. 11, p. 353, New York: Wiley 1978
749. Jansson, R. E. W. and Tomov, N. R.: Electrochim. Acta *25*, 497 (1980)
750. Ibl, N.: Electrochim. Acta *22*, 465 (1977)
751. Keller, R.: Electrochim. Acta *25*, 303 (1980)
752. Gallone, P.: Electrochim. Acta *22*, 913 (1977)
753. Beck, T. R. in: Yeager, E. and Salkind, A. J. (eds.): Techniques of Electrochemistry, Vol. 3, p. 1, New York: Wiley 1978
754. Lelandais, D.: Labo-Pharma.-Probl. Techn. *283*, 43 (1979)
755. Fioshin, M. Y.: Sov. Electrochem. *13*, 1 (1977)
756. Fleischmann, M. and Pletcher, D.: Chem. Brit. *11*, 50 (1975)
757. Grimshaw, J. in: Thirsk, H. R. (sen. rep.): Electrochemistry (Spec. Period. Rep.), Vol. 7, London: The Chemical Society 1980
758. Brown, O. R. in: Thirsk, H. R. (sen. rep.): Electrochemistry (Spec. Period. Rep.), Vol. 6, London: The Chemical Society 1978
759. Sawyer, D. T. and Roberts, Jr., J. L.: Experimental Electrochemistry for Chemists, New York: Wiley 1974
760. Pletcher, D.: Industrial Electrochemistry, London: Chapman and Hall 1982
761. Brown, O. R. in: Thirsk, H. R. (sen. rep.): Electrochemistry (Spec. Period. Rep.), Vol. 5, London: The Chemical Society 1975
762. Peover, M. E. in: Hush, N. S. (ed.): Reactions of Molecules at Electrodes, p. 259, London: Wiley 1971
763. Fleischmann, M. and Pletcher, D. in: Hush, N. S. (ed.): Reaction of Molecules at Electrodes, p. 347, London: Wiley 1971
764. Fleischmann, M. and Pletcher, D. in: Gold, V. (ed.): Advances in Physical Organic Chemistry, Vol. 10, London: Academic Press 1973
765. Prescott, J. H.: Chem. Eng. *72*, 238 (1965)
766. Baizer, M. M. (ed.): Organic Electrochemistry, New York: Dekker 1973
767. Baizer, M. M.: J. Appl. Electrochem. *10*, 285 (1980)
768. Beck, F.: Elektroorganische Chemie, Weinheim: Verlag Chemie 1973
769. Kyriacou, D. K.: Basis of Electroorganic Synthesis, New York: Wiley 1981
770. Schäfer, H. J.: Angew. Chem. *93*, 978 (1981)
771. Miller, L. L.: Pure Appl. Chem. *51*, 2125 (1979)
772. Eberson, L. and Weinberg, N. L.: Chem. Eng. News *49* (4), 40 (1971)
773. Fleischmann, M. and Pletcher, D.: Tetrahedron Lett. *1968*, 6255
774. Badoz-Lambling, J. and Cauquis, G. in: Nürnberg, H. W. (ed.): Electroanalytical Chemistry, London: Wiley 1974
775. Beck, F.: Chem. Ing. Techn. *42*, 153 (1970)
776. Hazelrigg, Jr., M. J. and Bard, A. J.: J. Electrochem. Soc. *122*, 211 (1975)
777. Ryan, M. D. and Evans, D. H.: J. Electroanal. Chem. Interfacial Electrochem. *67*, 333 (1976)
778. Cipris, D.: J. Electrochem. Soc. *127*, 1045 (1980)
779. Silvestri, G., Gambino, S., Filardo, G., Spadaro, S. and Palmisano, L.: Electrochim. Acta *23*, 413 (1978)
780. Braithwite, D. G.: U.S. Pat. 3,256,161 (1966)
781. Braithwite, D. G. and Bott, L. L.: U.S. Pat. 3,380,899 (1968)
782. Johnson, J. W. in: Bloom, H. and Gutmann, F. (eds.): Electrochemistry. The Past Thirty and the Next Thirty Years, New York: Plenum 1977
783. Kaiser, U. and Heitz, E.: Ber. Bunsenges. Phys. Chem. *77*, 818 (1973)
784. Skarlos, L.: U.S. Pat. 3,720,591 (1973)

785. Heitz, E. and Kaiser, U.: Ger. Offen. 2,301,032 (1974)
786. Amatore, C. and Savéant, J. M.: J. Amer. Chem. Soc. *103*, 5021 (1981)
787. Gambino, S. and Silvestri, G.: Tetrahedron Lett. *1973*, 3025
788. Silvestri, G., Gambino, S., Filardo, G., Guianazzi, M. and Ercoli, R.: Gazz. Chim. Ital. *102*, 818 (1972)
789. Cipris, D. and Mador, I. L.: J. Electrochem. Soc. *125*, 1954 (1978)
790. Knittel, D. and Kastening, B.: Ger. Offen. 2,328,196 (1974)
791. Knittel, D. and Kastening, B.: J. Appl. Electrochem. *3*, 291 (1973)
792. Knittel, D. and Kastening, B.: Ber. Bunsenges. Phys. Chem. *77*, 833 (1972)
793. Nyberg, K., Malmberg, M. and Servin, R.: in Ref. [733], p. 36
794. Bauer, R. and Wendt, H.: in Ref. [733], p. 56
795. Barbey, G., Delahaye, D., Lamant, M. and Caullet, C.: Electrochim. Acta *25*, 1273 (1980)
796. Lácan, M., Tabaković, I. and Čehović, Ž.: Tetrahedron *30*, 2911 (1974)
797. Miller, L. L., Kariv, E. and Behling, J. R. in: Clarke, F. H. (ed.): Annu. Rep. Med. Chem., Vol. 12, p. 309, New York: Academic Press 1977
798. Ronlan, A., Hammerich, O. and Parker, V. D.: J. Amer. Chem. Soc. *95*, 7132 (1973)
799. Falck, J. R., Miller, I. L. and Stermitz, F. R.: J. Amer. Chem. Soc. *96*, 2981 (1974)
800. Nohe, H.: in Ref. [733], p. 69
801. Jansson, R. E. W. and Fleischmann, M.: in Ref. [733], p. 2
802. Becker, J. Y., Byrd, L. R., Miller, L. L. and Ying-Huong, S.: J. Amer. Chem. Soc. *97*, 853 (1975)
803. Vincent, F., Tardivel, R. and Mison, P.: Tetrahedron Lett. *1975*, 603
804. Bockmair, G., Fritz, H. P. and Gebauer, H.: Electrochim. Acta *23*, 21 (1978)
805. Waefler, J. P. and Tissot, P.: Electrochim. Acta *23*, 899 (1978)
806. Koester, H. and Wendt, H. in: Bockris, J. O'M., Conway, B. E., Yeager, E. and White, R. A. (eds.): Comprehensive Treatise of Electrochemistry, Vol. 2, p. 251, New York: Plenum 1981
807. Nohe, H.: Chem.-Ing.-Techn. *46*, 594 (1974)
808. Pletcher, D. and Razaq, M.: J. Appl. Electrochem. *10*, 575 (1980)
809. Pletcher, D. and Razaq, M.: Electrochim. Acta *26*, 819 (1981)
810. Hatayama, T., Hamano, Y., Udo, K. and Yamamoto, T.: Japan. Pat. 7,441,192 (1974)
811. Ogumi, Z., Nishio, K. and Yoshizawa, S.: Electrochim. Acta *26*, 1779 (1981)
812. Merz, A.: Nachr. Chem. Techn. Lab. *30*, 16 (1982)
813. Parker, A. J.: Pure Appl. Chem. *53*, 1437 (1981)
814. Toušek, J.: Electrochim. Acta *22*, 47 (1977)
815. Gabe, D. R.: Metallography *5*, 415 (1972)
816. Breeze, P. A., Hartnagel, H. L. and Sherwood, P. M. A.: J. Electrochem. Soc. *127*, 454 (1980)
817. Hasegawa, H. and Hartnagel, H. L.: J. Electrochem. Soc. *123*, 713 (1976)
818. Müller, H., Eisen, F. H. and Mayer, J. W.: J. Electrochem. Soc. *122*, 651 (1975)
819. Turner, D. R. and Pankove, J. I. in: Yeager, E. and Salkind, A. J. (eds.): Techniques of Electrochemistry, Vol. 3, p. 105, New York: Wiley 1978
820. Binder, M., Gilman, S. and Wade, Jr., W.: J. Electrochem. Soc. *129*, 897 (1982)
821. Murray, R. W.: Acc. Chem. Res. *13*, 135 (1980)
822. Andrieux, C. P., Dumas-Bouchat, J. M. and Savéant, J. M.: J. Electroanal. Chem. Interfacial Electrochem. *131*, 1 (1982)
823. Snell, K. D. and Keenan, A. G.: Chem. Soc. Rev. *1979*, 259

XX References (Appendices A, B, C and D)

1. Riddick, J. A. and Bunger, W. B.: Organic Solvents in: Weissberger, A. (ed.), Techniques of Chemistry, Vol. 2, 3rd ed. New York: Wiley 1970
2. Wilhoit, R. C. and Zwolinski, B. J.: Physical and Thermodynamic Properties of Aliphatic Alcohols. New York: Amer. Chem. Soc. 1973
3. Alm, K. and Ciprian, M.: J. Chem. Eng. Data *25*, 100 (1980)
4. Srinivasan, K. R. and Kay, R. L.: J. Solution Chem. *4*, 299 (1975)
5. Cunningham, G. P., Vidulich, G. A. and Kay, R. L.: J. Chem. Eng. Data *12*, 336 (1967)

6. D'Aprano, A., Donato, I. D. and Caponetti, E.: J. Solution Chem. *3*, 371 (1974)
7. Barthel, J. et al.: unpublished data
8. Benson, G. C. and Kiyohara, O.: J. Solution Chem. *9*, 791 (1980)
9. Hales, J. L. and Ellender, J. H.: J. Chem. Thermodyn. *8*, 1177 (1976)
10. Brown, Jr., G. N. and Ziegler, W. T.: J. Chem. Eng. Data *24*, 319 (1979)
11. Ambrose, D. and Sprake, C. H. S.: J. Chem. Thermodyn. *2*, 631 (1970)
12. Kay, R. L. and Broadwater, T. L.: J. Solution Chem. *5*, 57 (1976)
13. Barthel, J., Schmeer, G., Strasser, F. and Wachter, R.: Rev. Chim. Miner. *15*, 99 (1978)
14. D'Aprano, A., Donato, I. D. and Caponetti, E.: J. Solution Chem. *8*, 135 (1979)
15. D'Aprano, A., Donato, I. D., Caponetti, E. and Agrigento, V.: J. Solution Chem. *8*, 793 (1979)
16. Evans, D. F., Thomas, J., Nadas, J. A. and Matesich, M. A.: J. Phys. Chem. *75*, 1714 (1971)
17. Dannhauser, W. and Bahe, L. W.: J. Chem. Phys. *40*, 3058 (1964)
18. Matesich, M. A., Nadas, J. A. and Evans, D. F.: J. Phys. Chem. *74*, 4568 (1970)
19. Evans, D. F. and McElroy, M. I.: J. Solution Chem. *4*, 413 (1975)
20. Evans, D. F. and Gardam, P.: J. Phys. Chem. *73*, 158 (1969)
21. Taniewska-Osińska, St. and Witkowski, St.: Acta Univ. Lodz, Ser. 2, *24*, 3 (1978)
22. Subrahmanyam, S. V. and Moorthy, N. M.: J. Solution Chem. *4*, 347 (1975)
23. Kalidas, C. and Sivaprasad, P. in: Furter, W. F. (ed.): Thermodynamic Behaviour of Electrolytes in Mixed Solvents, Vol. 2, p. 345, Washington: American Chemical Society 1979
24. DeSieno, R. P., Greco, P. W. and Mamajek, R. C.: J. Phys. Chem. *75*, 1722 (1971)
25. Hayduk, W. and Malik, V. K.: J. Chem. Eng. Data *16*, 143 (1971)
26. Douhéret, G. and Morénas, M.: Can. J. Chem. *57*, 608 (1979)
27. Blokhra, R. L. and Sehgal, Y. P.: J. Solution Chem. *5*, 399 (1976)
28. Merken, G. V., Thun, H. P. and Verbeek, F.: Electrochim. Acta *21*, 11 (1976)
29. Barker, B. J., Rosenfarb, J. and Mullin, T.: J. Solution Chem. *6*, 513 (1977)
30. Dulić, N. and Horváth, L.: Acta Phys. Chem. *24*, 451 (1978)
31. Abraham, T., Bery, V. and Kudchadker, P.: J. Chem. Eng. Data *16*, 355 (1971)
32. Vitagliano, V., Zagari, A. and Sartorio, R.: J. Chem. Eng. Data *18*, 370 (1973)
33. Hafez, M. and Hartland, S.: J. Chem. Eng. Data *21*, 179 (1976)
34. Ratkovics, F., Salamon, T. and László, M.: Acta Chim. Acad. Sci. Hung. *66*, 259 (1970)
35. Thomas, J. and Evans, D. F.: J. Phys. Chem. *74*, 3812 (1970)
36. Della Monica, M. and Bufo, S.: Electrochim. Acta *22*, 1213 (1977)
37. Bruno, P. and Della Monica, M.: Electrochim. Acta *20*, 179 (1975)
38. Reid, D. S. and Vincent, C. A.: J. Electroanal. Chem. *18*, 427 (1968)
39. Kozlowski, Z., Kinart, C. and Bald, A.: Rocz. Chem. *51*, 1471 (1977)
40. Dack, M. R. J.: Aust. J. Chem. *28*, 1643 (1975)
41. Hamilton, D. and Stokes, R. H.: J. Solution Chem. *1*, 213 (1972)
42. Rohdewald, P. and Moldner, M.: J. Phys. Chem. *77*, 373 (1973)
43. de Visser, C., Pel, P. and Somsen, G.: J. Solution Chem. *6*, 571 (1977)
44. Kreis, R. W. and Wood, R. H.: J. Chem. Thermodyn. *1*, 523 (1969)
45. Kortüm, G. and Hebestreit, Ch.: Z. Phys. Chem. N.F. *93*, 235 (1974)
46. Thompson, P. T., Durbano, M., Turner, J. L. and Wood, R. H.: J. Solution Chem. *9*, 955 (1980)
47. Moellmer, J. F., Ott, J. B., Goates, J. R. and Farrell, D. W.: J. Chem. Eng. Data *21*, 317 (1976)
48. Aoyagi, K. and Albright, J. G.: J. Solution Chem. *6*, 635 (1977)
49. Atlani, Ch. and Justice, J.-C.: J. Solution Chem. *4*, 955 (1975)
50. Bollinger, J.-C., Yvernault, G. and Yvernault, Th.: J. Solution Chem. *7*, 317 (1978)
51. Hanna, E. M., Pethybridge, A. D., Prue, J. E. and Spiers, D. J.: J. Solution Chem. *3*, 563 (1974)
52. Sacco, A., Petrella, G., Della Monica, M. and Castagnolo, M.: J. Chem. Soc., Faraday Trans. 1 *73*, 1936 (1977)
53. Okpala, Ch., Guiseppi-Elie, A. and Maharajh, D. M.: J. Chem. Eng. Data *25*, 384 (1980)
54. Yao, N.-P. and Bennion, D. N.: J. Electrochem. Soc. *118*, 1097 (1971)
55. Schiavo, S., Fuoss, R. M., Marrosu, G. and Guida, G.: J. Solution Chem. *8*, 557 (1979)
56. Casteel, J. F. and Sears, P. G.: J. Chem. Eng. Data *19*, 196 (1974)
57. Cooke, C., McCallum, C., Pethybridge, A. D. and Prue, J. E.: Electrochim. Acta *20*, 591 (1975)

58. Bicknell, R. T. M., Lawrence, K. G. and Feakins, D.: J. Chem. Soc. Faraday Trans. 1, *76*, 637 (1980)
59. D'Aprano, A.: J. Solution Chem. *3*, 363 (1974)
60. Covington, A. K. and Dickinson, T. (eds.): Physical Chemistry of Organic Solvent Systems, p. 5, London: Plenum 1973
61. Srinivasan, K. R. and Kay, R. L.: J. Solution Chem. *6*, 357 (1977)
62. Akhmetkarimov, K. A., Mai, I. I. and Muldakhmetov, Z. M.: J. General. Chem. USSR (Engl. Transl.) *43*, 460 (1973)
63. Hafez, M. and Hartland, S.: J. Chem. Eng. Data *21*, 179 (1976)
64. Niki, E., Kamiya, Y. and Ohta, N.: Bull. Chem. Soc. Jpn. *42*, 3578 (1969)
65. Shkodin, A. M., Podolyanko, V. A. and Mikhailova, E. N.: Sov. Electrochem. (Engl. Transl.) *10*, 399 (1974)
66. Pirson, D. J. and Huyskens, P. L.: J. Solution Chem. *3*, 503 (1974)
67. Aelenei, N.: Bull. Inst. Politeh. Iasi, Sect. 2: Chim. Ing. Chim. *25*, 41 (1979)
68. Miller, L. P., Wachter, H. N. and Fried, V.: J. Chem. Eng. Data *20*, 417 (1975)
69. Petrella, G., Castagnolo, M., Sacco, A. and De Giglio, A.: J. Solution Chem. *5*, 621 (1976)
70. Ashcroft, St. J., Clayton, A. D. and Shearn, R. B.: J. Chem. Eng. Data *24*, 195 (1979)
71. Sears, P. G., Stoeckinger, Th. M. and Dawson, L. R.: J. Chem. Eng. Data *16*, 220 (1971)
72. D'Aprano, A.: Gazz. Chim. Ital. *104*, 91 (1974)
73. Boerner, B. R. and Bates, R. G.: J. Solution Chem. *7*, 245 (1978)
74. Petrella, G. and Sacco, A.: J. Chem. Soc., Faraday Trans. 1 *74*, 2070 (1978)
75. Jones, A. R. and Aikens, D. A.: J. Chem. Eng. Data *27*, 24 (1982)
76. Payne, R. and Theodorou, I. E.: J. Phys. Chem. *76*, 2892 (1972)
77. Jansen, M. L. and Yeager, H. L.: J. Phys. Chem. *77*, 3089 (1973)
78. Werblan, L. and Lesiński, J.: Pol. J. Chem. *54*, 507 (1980)
79. Islam, N., Islam, M. R. and Ahmad, M.: Indian J. Chem. *17A*, 126 (1979)
80. Tommila, E. and Lindell, E.: Suom. Kemistil. *B42*, 93 (1969)
81. Tommila, E. and Yrjövuori, R.: Suom. Kemistil. *B42*, 90 (1969)
82. Mussche, M. J. and Verhoeye, L. A.: J. Chem. Eng. Data *20*, 46 (1975)
83. Kay, R. L. and Broadwater, T. L.: Electrochim. Acta *16*, 667 (1971)
84. Fratiello, A. and Kay, R. L.: J. Solution Chem. *3*, 857 (1974)
85. Scott, D. W.: J. Chem. Thermodyn. *2*, 833 (1970)
86. Hogen-Esch, T. E. and Smid, J.: J. Amer. Chem. Soc. *88*, 318 (1966)
87. Carvajal, C., Tölle, K. J., Smid, J. and Szwarc, M.: J. Amer. Chem. Soc. *87*, 5548 (1965)
88. Comyn, J., Dainton, F. S. and Ivin, K. J.: Electrochim. Acta *13*, 1851 (1968)
89. Hayduk, W., Laudie, H. and Smith, O. H.: J. Chem. Eng. Data *18*, 373 (1973)
90. Jagodzinski, P. and Petrucci, S.: J. Phys. Chem. *78*, 917 (1974)
91. Kusano, K.: J. Chem. Eng. Data *23*, 141 (1978)
92. Renard, E. and Justice, J.-C.: J. Solution Chem. *3*, 633 (1974)
93. Takenaka, N. and Arakawa, K.: Bull. Chem. Soc. Jpn. *47*, 566 (1974)
94. Canters, G. W.: J. Amer. Chem. Soc. *94*, 5230 (1972)
95. Phillips, T. W. and Murphy, K. P.: J. Chem. Eng. Data *15*, 304 (1970)
96. Wade, J. C. and Taylor, Jr., Z. L.: J. Chem. Eng. Data *18*, 424 (1973)
97. Badiali, J.-P., Cachet, J., Cyrot, A. and Lestrade, J.-C.: J. Chem. Soc., Faraday Trans. 2 *69*, 1339 (1973)
98. Tommila, E. and Autio, T.: Soum. Kemistil. *B42*, 107 (1969)
99. Paljk, S. and Klofutar, C.: J. Chem. Soc., Faraday Trans. 1 *74*, 2159 (1978)
100. Hales, J. L. and Townsend, R.: J. Chem. Thermodyn. *4*, 763 (1972)
101. Fernández-Prini, R. and Urrutia, G.: J. Chem. Soc., Faraday Trans. 1 *72*, 637 (1976)
102. Kiyohara, O. and Benson, G.: J. Chem. Eng. Data *26*, 263 (1981)
103. Kolthoff, I. M. in: Marchon, J. C. (ed.): Non Aqueous Electrochemistry, p. 312, London: Butterworths 1971
104. Dack, M. R. J. in: Solutions and Solubilities, Part II, in: Weissberger, A. (ser. ed.): Techniques of Chemistry, Vol. 8, p. 95, New York: Wiley 1976
105. Assarsson, P. and Eirich, F. R.: J. Phys. Chem. *72*, 2710 (1968)
106. Ioffe, B. V.: Zh. Obsh. Khim. *25*, 902 (1955)
107. Barker, B. J. and Caruso, J. A.: J. Amer. Chem. Soc. *93*, 1341 (1971)

108. Seddon, W. A., Fletcher, J. W., Sopchyshyn, F. C. and Caterall, R.: Can. J. Chem. *55*, 3356 (1977)
109. D'Aprano, A., Komiyama, J. and Fuoss, R. M.: J. Solution Chem. *5*, 279 (1976)
110. Gutmann, V.: The Donor-Acceptor Approach to Molecular Interactions, New York: Plenum 1978
111. Reichardt, C.: Solvent Effects in Organic Chemistry, Weinheim: Verlag Chemie 1979
112. Mayer, U., Gutmann, V. and Gerger, W.: Monatsh. Chem. *106*, 1235 (1975)
113. Mayer, U.: Coord. Chem. Rev. *21*, 159 (1976)
114. Marcus, Y.: Introduction to Liquid State Chemistry, London: Wiley 1977
115. Barthel, J., Popp, H. and Schmeer, G.: Data bank ELDAR (*El*ectrolyte *Da*ta Bank *R*egensburg), DECHEMA project
116. Robinson, R. A. and Stokes, R. H.: Electrolyte Solutions, London: Butterworths, 2nd rev. ed. 1970
117. Goldschmidt, V. M.: Chem. Ber. *60*, 1263 (1927)
118. Grunwald, E., Baughman, G. and Kohnstamm, G.: J. Amer. Chem. Soc. *82*, 5801 (1960)
119. Gourary, B. S. and Adrian, F. J.: Solid State Phys. *10*, 127 (1960)
120. Ladd, M. F. C.: Theor. Chim. Acta *12*, 333 (1968)
121. Jenkins, H. D. B. and Thakur, K. P.: J. Chem. Educ. *56*, 576 (1979)
122. Conway, B. E.: Ionic Hydration in Chemistry and Biophysics, Amsterdam: Elsevier 1981
123. Krumgalz, B. S.: J. Chem. Soc., Faraday Trans. 1 *78*, 437 (1982)
124. Barthel, J. in: European Federation of Chemical Engineering (ed.): Phase Equilibria and Fluid Properties in the Chemical Industry, Vol. 2, p. 497, Frankfurt a. M.: DECHEMA 1980
125. Mitchell, A. D. and Cross, L. C. (eds.): Tables of Interatomic Distances and Configuration in Molecules and Ions, Spec. Publ. No. 11, London: The Chemical Society 1958
126. Bondi, A.: Physical Properties of Molecular Crystals, Liquids and Glasses, New York: Wiley 1968
127. Barthel, J.: Habilitationsschrift, Saarbrücken 1958
128. Schmeer, G.: Habilitationsschrift, Regensburg 1981
129. Böttcher, C. J. F.: Theory of Electric Polarisation Amsterdam: Elsevier 1952
130. Mayer, U. in: Yeager, E. B., Schumm jr., B., Blomgren, G., Blankenship, D. R., Leger, V. and Akridge, J. (eds.): Proc. Workshop on Lithium Nonaqueous Battery Electrochemistry, p. 13, Pennington, N. J.: The Electrochem. Soc. 1980
131. Compiled by Cohen, E. R. and Taylor, B. N.; officially adopted by CODATA; published in: J. Phys. Chem. Ref. Data, Vol. 2, No. 4, 663 (1973); CODATA Bulletin No. 11 (Dec. 1973); Dimensions/NBS (Jan. 1974)

Subhalides of Tellurium

Rüdiger Kniep[1] **and Albrecht Rabenau**[2]

[1] Institut für Anorganische Chemie und Strukturchemie, Universität Düsseldorf, Universitätsstr. 1, D-4000 Düsseldorf, FRG

[2] Max-Planck-Institut für Festkörperforschung, Heisenbergstr. 1, D-7000 Stuttgart 80, FRG

Table of Contents

The tellurium subhalides represent a group of recently discovered compounds which have been investigated by the methods of solid-state and structural chemistry. Because of their high tellurium content and their crystal structures containing homonuclear tellurium connections they have attained significance as "modified tellurium structures". As a consequence, tellurium subhalides have been the subject of various investigations in the fields of solid-state and chemical physics.
In this paper a short survey of the group of tellurium halides is followed by detailed information on phase relations, preparation, crystal growth, and crystal structures of tellurium subhalides. Mössbauer investigations as well as optical, photoelectric, and electrical properties of the respective compounds are discussed. Solid-state galvanic cells containing tellurium subhalides have enabled the precise determination of relevant thermodynamic data.

I Introduction

Tellurium halides have been the subject of chemical research since Rose [1] (1831) and Berzelius [2] (1835). From the point of view of classical inorganic chemistry it is easy to accept the fact that compounds with the compositions TeX_6, TeX_4 and TeX_2 [1] were predominantly under discussion at that time. Nevertheless, early thermoanalytical investigations in the binary Te—X systems [3,4,5] led to the conclusion that the compounds TeX_6 do not exist, and no contradiction to this statement has appeared up to now. There is no doubt concerning the existence of solid tetrahalides (I.A). The dihalides, on the other hand, only occur in the vapour state although methods of preparation of the respective solid compounds were reported in the literature up to the middle seventies (I.B). Tellurium subhalides represent a rather recently discovered group within the halides of tellurium. Their general formula Te_yX ($y \geqq 1$) indicates a high tellurium content which has aroused special interest in the fields of chemical and physical research.

I.A Tetrahalides

Tellurium tetrahalides ($TeCl_4$, $TeBr_4$, TeI_4) are well-known crystalline compounds with melting temperatures (congruent) 223 °C [5,6,7]; 388 °C [7,8] and 280 °C [4,9,10,11] respectively. Crystal structure analyses have been carried out on $TeCl_4$ [12,13] and TeI_4 [14]. $TeBr_4$ is an isotype of the tetrachloride [13,15,16]. Crystallographic data of the tetrahalides are summarized in Table 1. In addition, there are some hints in the literature concerning the existence of at least more than one crystalline phase of composition TeI_4: In the course of cooling TeI_4-melts down to room temperature, a crystalline tetragonal phase (a = 1612(4), c = 1120(2) pm) was obtained and reported to have the composition TeI_4 [17]. In a more recent work [18] crystalline TeI_4 (composition checked by chemical analysis) with an X-ray powder pattern different from that of the corresponding diagram of the modification in Table 1 was obtained by transport of the vapour in contact with a 2:1 mixture of tellurium and iodine, using argon as a carrier gas and transport temperatures from 190 °C to 65 °C.

The crystal structure of $TeCl_4$ [12,13] consists of isolated tetramers $(TeCl_4)_4$ which have a cubane-like shape with the tellurium and chlorine atoms occupying alternately the corners of a Te_4Cl_4 skeleton; the remaining chlorine atoms are situated in terminal positions, three for every tellurium. In the TeI_4 structure [14] there are also isolated $(TeI_4)_4$ tetramers, but in a different arrangement: four TeI_6-octahedra share common edges such that four of the 16 iodine atoms are bound to two tellurium atoms and two iodine atoms are bound to three tellurium atoms, the remaining 10 iodines being in terminal positions.

$TeCl_4$ and $TeBr_4$ form a continuous series of mixed crystals with a two-phase region (solid solution + melt) situated between the melting temperatures of the pure components [10,19]. Furthermore, it has been assumed [10] that similarly the

[1] X = chlorine, bromine and iodine respectively; fluorine compounds are not under consideration in this review.

Table 1. Crystallographic data for tellurium tetrahalides.

Compound	Symmetry Space Group	a	b β (°)	c (pm)	Z	D_x ($g \cdot cm^{-3}$)	Ref.
$TeCl_4$	mcl C2/c	1707.6 (5)	1040.4 (5) 116.82 (5)	1525.2 (8)	16	2.595	[12,13] [15]
$TeBr_4$	mcl C2/c	1775 (3)	1089 (2) 116.6 (2)	1588 (2)	16	4.236	[see also [16]]
TeI_4	orh Pnma	1363.5 (5)	1679.8 (5)	1462.5 (5)	16	5.039	[14]

systems $TeCl_4$—TeI_4 and $TeBr_4$—TeI_4 contain continuous series of solid solutions. This however does not correspond to the fact that the pure components of the respective quasibinary systems are of different structural types. Actually, a reinvestigation of the phase relations in the systems $TeCl_4$—TeI_4 and $TeBr_4$—TeI_4 [19] showed them to exhibit simple eutectics with only limited regions of solid solutions.

I.B Dihalides

Phase diagram studies of the quasibinary systems Te—$TeCl_4$ [5,6,7,20], Te—$TeBr_4$ [5,7,8] and Te—TeI_4 [3,4,5,9,11,21] led to the result that no dihalide exists in the solid state under conditions of thermodynamic equilibrium. Therefore, the question remained whether the solid dihalides, reported by a number of different authors, have been obtained under non-equilibrium conditions or whether they are actually composed of two or more different phases.

The preparation of solid $TeCl_2$ and investigations on this compound are reported in [1,2,22,23,24,25,26,27]. It was shown by thermoanalytical investigations [5,28] that the products obtained by reaction of the elements followed by distillation [1,2] consist of more than one phase. Preparation of $TeCl_2$ by reaction of tellurium with CF_2Cl_2 [23] was reinvestigated [29] and reported to yield a mixture of $TeCl_4$ and probably amorphous tellurium. Commercial $TeCl_2$ has been used for magnetic [22] and spectroscopic [27] investigations without checking the compound by chemical analysis. A Mössbauer study on $TeCl_2$ [26] at liquid helium temperature was carried out without further characterization of the material under investigation. $TeCl_2$ was reported to be formed by the reaction $SnCl_2 + TeCl_4 \rightarrow SnCl_4 + TeCl_2$ [24,25]; a reinvestigation however showed $TeCl_4$ to be reduced to elemental tellurium ($2\,SnCl_2 + TeCl_4 \rightarrow Te + 2\,SnCl_4$) [30]. It may be stated in this connection that the phase diagram $TeCl_2$—KCl is not quasibinary with an intermediate phase $TeCl_2 \cdot KCl$ [25] but pseudobinary [31], and the ternary phase is of composition K_2TeCl_6.

Preparation of solid $TeBr_2$ and investigations of this compound are reported in [2,5,22,26,32,33]. The products obtained by evaporation of Te/$TeBr_4$ mixtures and quenching of the vapour [2,5,33] are described as brown solids which are unstable and transform to a mixture of different phases even on gentle heating. A similar material was prepared by reaction of tellurium with CF_3Br [32]. Commercial $TeBr_2$ was used for magnetic investigations [22], and Mössbauer measurements [26] have been carried out without further characterization of the material under investigation.

With regard to solid TeI_2 there is only one description of the preparation [2] which starts from the elements. In the course of a thermoanalytical investigation of the Te—I system [3] this material [2] was shown to consist of more than one phase. Commercial TeI_2 was used [34] for a spectroscopic study of the corresponding vapour phase without characterization of the starting material. After all, it is amazing to note that a Mössbauer study on TeI_2 at liquid helium temperature [26] led to the conclusion that solid TeI_2 is a definite molecular phase with a geometry of the molecule being angular (I—Te—I = 150°).

It can be seen from the preceding statements that tellurium dihalides do not exist in the solid state. The materials under investigations, if actually binary, with great certainty consisted of more than one phase. On the other hand, there is no doubt about the existence of dihalide molecules in the vapour phase. This is confirmed by a number of investigations on the vapour pressure in the respective systems [27,35,36,37,38], by experiments concerning the chemical transport of tellurium with iodine [39] and by electron diffraction studies [33,40] (II.D).

I.C Subhalides — General Considerations

In 1956, the first tellurium subhalide (characterized as Te_nI_n) was found in X-ray investigations of the Te—TeI_4 system [17]. About 10 years later, an additional subiodide (Te_xI) [41] was prepared by hydrothermal synthesis [42]. Systematic reinvestigations of the phase relations in the quasibinary systems Te—$TeCl_4$ [6,7], Te—$TeBr_4$ [7,8] and Te—TeI_4 [9,21] were undertaken in order to clear up the actual existence and variety of subhalides of tellurium. The following crystal structure analyses showed these compounds to be appropriately regarded as "modified tellurium structures". This finding was consequently followed by extensive research on the physical properties of the subhalides. The close collaboration of solid-state chemistry and solid-state physics provided extensive knowledge of this group of compounds within a period of little more than one decade.

II Phase Relations

II.A Phase Diagrams

The knowledge of phase diagrams and phase relations is of fundamental significance in the field of solid-state chemistry. The actual existence and variety of intermediate stable phases as well as the p-T-x behaviour of a particular phase are illustrated by a phase diagram. Thus, phase diagrams provide a basis for considering the optimum conditions for crystal growth which in turn provide a basis concerning optimum investigations of physical properties of crystalline materials.

II.A.1 Te—$TeCl_4$

The system Te—$TeCl_4$ has been investigated by thermoanalytical and X-ray analytical methods [5,6,7,20,43]. The final result [7] is given in Fig. 1. The system is

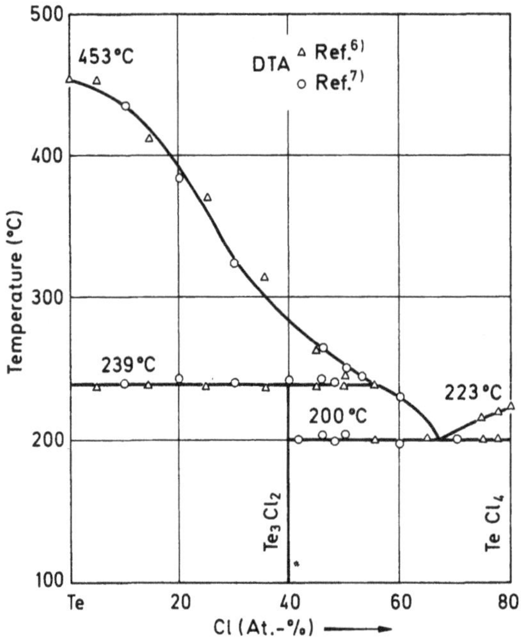

Fig. 1. Phase diagram Te-$TeCl_4$. Though there are differences in the composition of the intermediate compound (see text), liquidus and solidus temperatures of the investigations [6] and [7] agree very well

quasibinary and contains one intermediate compound of composition Te_3Cl_2 which melts incongruently at 239 °C. The existence and composition of this subchloride have been confirmed by determination of its crystal structure [44,45,46].

The occurrence of the intermediate phase Te_3Cl_2 [7] was not observed in [5,6,20,43]; the system was reported to be of a simple eutectic type [6,43], eutectic with regions of solid solution [5] and peritectic with an intermediate phase TeCl [6], respectively. The interpretation of the system as simple eutectic was based only on DTA measurements during cooling [43], and no details of the procedure of investigation were given. Determination of the Tamman triangles in the Te—$TeCl_4$ system [5,6] yielded two maxima, one maximum being in agreement with the eutectic composition given in Fig. 1 and the other at a composition of 50 atom-% Cl. The interpretation of the latter maximum as referring to an intermediate phase TeCl [6] is not correct because the eutectic line at 200 °C extends from $TeCl_4$ to Te_3Cl_2. As can be seen from Fig. 1 no regions of solid solution [5] are present; it may be stated in this connection, that experimental data and interpretation in [5] are inconsistent and do not obey to the phase rule.

II.A.2 Te—$TeBr_4$

The system Te—$TeBr_4$ has been investigated by thermoanalytical, X-ray analytical and metallographic methods [5,7,8]. The final result [7,8] is given in Fig. 2. The system is quasibinary and contains one intermediate compound of composition Te_2Br which melts incongruently at 225 °C. The existence and composition of this subbromide have been confirmed by determination of its crystal structure [44,45,46].

In contrast to the phase diagram given in Fig. 2, the system has been reported [5]

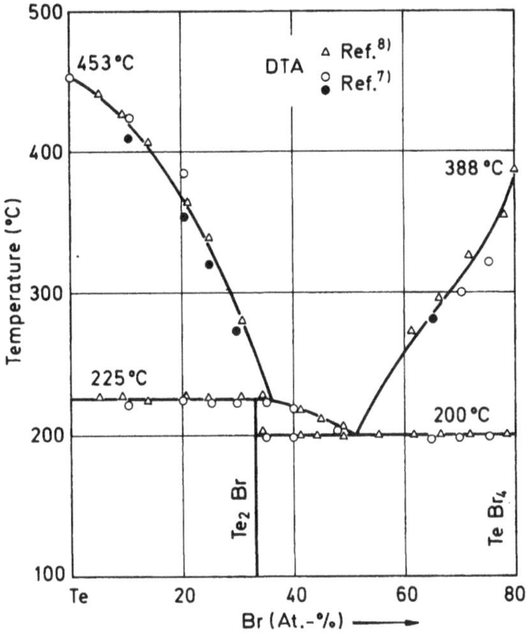

Fig. 2. Phase diagram Te-$TeBr_4$ [7,8]

to be eutectic with regions of solid solutions. It is interesting to note that the determination of the Tamman triangles in the system Te—$TeBr_4$ led to corresponding results [5,8] but different interpretations. Comparison of the composition of the Tamman maxima with X-ray powder analyses [8] led to the conclusion that one maximum corresponds to the eutectic composition, the other one to the composition of the intermediate phase Te_2Br. The finding of solid solutions [5] instead of the formation of Te_2Br may be connected with the accompanying metallographic investigations. It seems to be reasonable that polishing and etching of respective samples caused decomposition of Te_2Br as well as $TeBr_4$ with formation of metallic tellurium which was then detected. Moreover, experimental data and interpretation in [5] do not follow the phase rule.

II.A.3 Te—TeI_4

The system Te—TeI_4 has been investigated by thermoanalytical, X-ray analytical, metallographic and vapour pressure methods [3,4,5,9,11,17,21]. The results [9] are illustrated by Fig. 3; liquidus temperatures, to some extent, are based on vapour pressure data shown in Figs. 4 and 5. The system is quasibinary and contains one intermediate compound of composition TeI which melts incongruently at 185 °C. The existence and composition of this subiodide have been confirmed by determination of its crystal structure [44,45,47].

In contrast to the phase diagram given in Fig. 3 the system has been reported to be simple eutectic [3], and eutectic with regions of solid solutions [4,5], respectively. It may be noted that thermoanalytical investigations involving cooling from the melt [3] do not provide conditions of thermodynamic equilibrium and furthermore

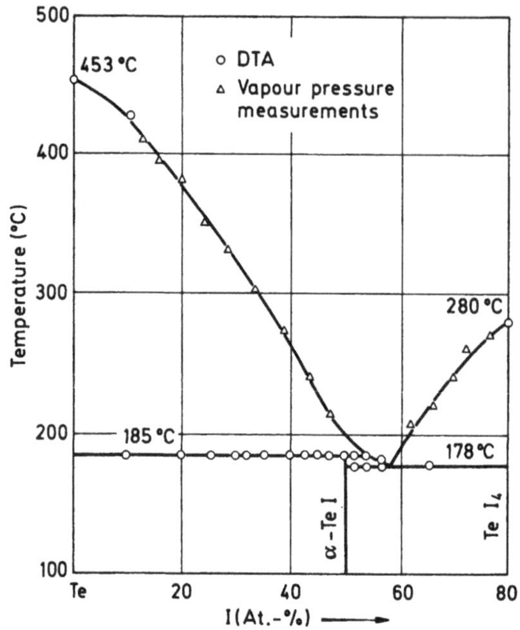

Fig. 3. Phase diagram Te-TeI_4 [9]; in general, there is agreement with a study of the system [21] which found the eutectic and peritectic temperatures about 5 °C below the given temperatures. The notation α for the intermediate subiodide corresponds to the existence of a metastable β-phase of the same composition (II.B)

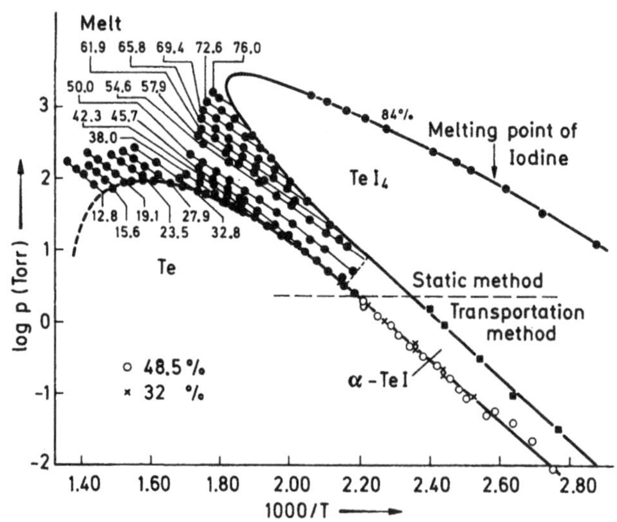

Fig. 4. Vapour pressure curves of tellurium-iodine mixtures [9]. The parameter is the iodine content in atom-% of the condensed phases. The two liquidus branches below 45.7 atom-% I and above 57.9 atom-% I belong to the equilibria Te + liquid and TeI_4 + liquid, respectively. The part inbetween (dotted line) corresponds to the equilibrium α-TeI + melt

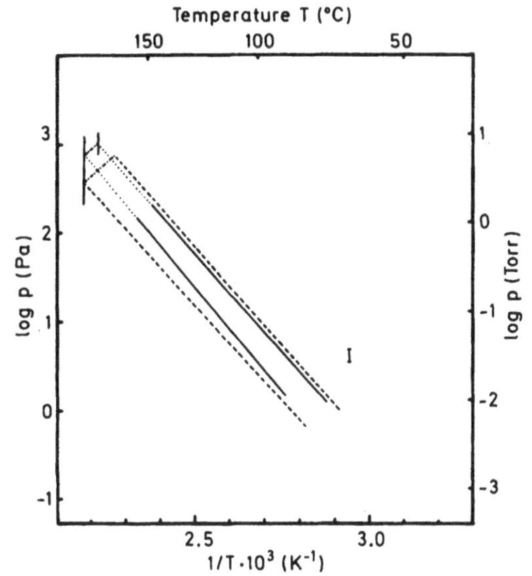

Fig. 5. Vapour pressure over α-TeI-Te (lower two curves) and α-TeI-TeI_4 (upper two curves) mixtures. ———, calculated from e.m.f. measurements using respective galvanic cells[95] (see IX.D) [........, extrapolation to higher temperatures]; — — — —, calculated[103] from results obtained by using the transportation method; —·—·—·, equilibrium between α-TeI and the melt. The error bar refers to the data given in[103]. The two vertical lines indicate the peritectic and the eutectic temperatures

give rise to problems from the effect of supercooling. A metallographic study[4,5] of samples of the Te—TeI_4 system seems to be disadvantageous because polishing and etching may cause decomposition of intermediate phases with evolution of elemental tellurium which is then detected. Moreover, experimental data and interpretation in[4,5] are not in agreement with the phase rule.

II.B Metastable Crystalline Phases

As shown in Section II.A, the intermediate subhalides Te_3Cl_2, Te_2Br and α-TeI are stable phases. In addition, a number of metastable subhalides of tellurium have been found.

In the course of the hydrothermal growth[42,48] of tellurium and tellurium iodides[41,44,49,50] two metastable subiodides, Te_2I and β-TeI, have been obtained. β-TeI is identical with Te_xI[11,41]; the supposed range of composition of Te_xI was attributed to the simultaneous presence of Te_2I the crystals of which are formed together with β-TeI in varying amounts and with a similar shape[9,44]. Te_2I crystals decompose on treatment in a mortar and are therefore not observed amongst other phases in the corresponding X-ray powder patterns. The metastable character of Te_2I and β-TeI was first shown[44,47] by annealing mixtures of the phases at 170 °C (this is 15 °C below the peritectic temperature of the Te—TeI_4 system; see Fig. 3) for ten days followed by X-ray powder investigation. The powder patterns were found to consist exclusively of the stable phase combination Te + α-TeI. Furthermore, the phase transformation Te_2I/β-TeI → Te/α-TeI, which occurred in all experiments performed, was shown to be of a monotrope type. The only observation of a transformation α-TeI → β-TeI[11] was due to partial fusion of the sample at 190 °C[9]; in correspondence with Oswald's rule the phase which is formed by cooling of respective melts is metastable β-TeI. The actual existence and composition of the metastable subiodides have been confirmed by determination of their crystal structures [44,45,47].

An additional metastable subchloride, Te_2Cl, was first reported in [18]. The only method for the preparation of Te_2Cl crystals is the condensation of the vapour in contact with a Te/Cl melt of 1:1 composition. Therefore, because of the nature of the phase diagram Te—$TeCl_4$ (see Fig. 1), which contains only Te_3Cl_2 as an intermediate phase, Te_2Cl is regarded as a metastable compound. The actual existence and composition of Te_2Cl have been confirmed by X-ray investigations [44,45] which showed Te_2Cl to be an isotype of Te_2Br and Te_2I.

II.C Liquid Phases

It can be seen from Figs. 1–3 that no regions of liquid immiscibility appear in the melt ranges of the Te—TeX_4 (X = Cl, Br, I) systems. Thermodynamic calculations [51] suggested that $TeCl_2$- and $TeBr_2$-species occur, to a certain extent, within homogeneous melts of the respective systems at elevated temperatures; no evidence of the existence of these species within the two-phase regions solid + liquid has been reported.

Measurements of the electrical conductivity of molten Te + I_2 solutions [52] indicate the presence of three different types of melts in the Te—TeI_4 region: a) Te_{100} — $Te_{90}I_{10}$ with a conductivity versus temperature plot similar to pure molten tellurium; b) $Te_{90}I_{10}$ — $Te_{50}I_{50}$ characterized by a continuous decrease of the electrical conductivity and c) $Te_{50}I_{50}$ — $Te_{20}I_{80}$ with a constant low electrical conductivity. The change of electrical conductivity to lower values with increasing iodine content is assumed to be connected with a decrease of the relative size of homonuclear Te-fragments which are present in the respective melts. This assumption is largely consistent with X-ray investigations on the solid-state compositions obtained after cooling Te/TeI_4 melts of varying ratios down to room temperature without annealing [18,50]. In the range Te_{100} — $Te_{68}I_{32}$ only elemental tellurium (3_1 chain structure [53]) and β-TeI (1/∞ homonuclear Te-chains [44,47]) were observed whereas in the range $Te_{68}I_{32}$ — $Te_{50}I_{50}$ the solid-phase composition Te + β-TeI (decreasing amounts) + α-TeI (increasing amounts) was obtained. The crystal structure of α-TeI is a molecular structure containing only a homonuclear Te_4-ring [44,47]. On cooling $Te_{50}I_{50}$ — — $Te_{20}I_{80}$ melts a phase combination of the molecular phases α-TeI and TeI_4 was obtained.

Cooling of $Te_2Br_{1-x}I_x$ melts with the composition $0 \geqq x \leqq 0.75$ down to room temperature yielded vitreous metastable phases [18]. The vitreous phases and their short-range orderes have been the subject of X-ray diffraction studies [54].

These studies revealed that the homonuclear structure of condensed Te-chains which is present in the crystalline form [44,45] also exists in the vitreous phase with an extension of ~1600 pm. Considering the vitreous phase to be a supercooled melt, similar structural units are expected to occur in the respective melts as well.

II.D Vapour Phases

The vapour of tellurium halide systems and the molecular species in the gaseous state have been the subject of a number of investigations. This is partly due to the fact that the tellurium halides are succesfully used for the chemical transport of chalcogenides [36].

Tellurium chloride [6,36,55] and bromide [36,55,56] systems contain, in addition to the corresponding tetrahalides, dihalide species in the vapour phase. They are formed by thermal decomposition of the tetrahalide (1):

$$TeX_{4(g)} \rightleftharpoons TeX_{2(g)} + X_{2(g)} \qquad (1)$$

The relative amount of TeX_2 in the vapour increases with rising temperature, and the decomposition of TeX_4 is nearly complete at temperatures $\geqq 500$ °C. $TeCl_2$ and $TeBr_2$ in the vapour were investigated by electron diffraction [33,40] and also characterized by their electronic absorption spectra [27,57]. Monohalide species (TeX) have been observed by flash photolyses of the vapour and by recording the UV absorption spectra [58,59].

There has been some confusion concerning the existence of TeI_2 in the vapour phase. An early vapour pressure determination [56] of the Te—I system resulted in the statement that the only binary molecular species in the gaseous state is TeI_4. On the other hand, in the course of the chemical transport of tellurium with iodine [39] (2) it was shown by mass spectroscopy [60] that the main species in the gas phase is TeI_2 if a tellurium-rich binary system is heated under vacuum.

$$Te_{(s)} + I_{2(g)} \rightleftharpoons TeI_{2(g)} \qquad (2)$$

The following vapour pressure investigations confirmed the presence of TeI_4 molecules in the vapour [38,55,61,62,63]. Different interpretations of the existence of an additional binary component, which was supposed to be $TeI_{(g)}$ [61] or $TeI_{2(g)}$ [38] were reported. The decomposition of TeI_4 according to reaction (3) without formation of a further binary molecular compound was also claimed. [55,62,63].

$$TeI_{4(g)} \rightleftharpoons Te_{(s)} + 2\,I_{2(g)} \qquad (3)$$

Recent investigations on the vapour pressure above TeI_4 [37,64] have shown that in addition to reaction (1) reactions (2) and (3) may contribute to the observed total pressure and its temperature dependence up to 600 °C. At temperatures >600 °C additional thermal decomposition of I_2 (4) and TeI_2 (5) takes place.

$$I_{2(g)} \rightleftharpoons 2\,I_{(g)} \qquad (4)$$

$$TeI_{2(g)} \rightleftharpoons Te_{2(g)} + 2\,I_{2(g)} \qquad (5)$$

A monoiodide species TeI in the gas phase has been observed by flash photolysis of TeI_2 vapour and by recording of the UV absorption spectrum [58,65].

III Preparation and Crystal Growth

Tellurium subhalides only exist in the solid state and are not dissolved by usual chemical solvents without irreversible decomposition. There is no chemical method available which separates mixtures containing tellurium and tellurium subhalides. On the other hand, chemical extraction of solid tetrahalides from mixtures of tetrahalides and subhalides is known to be practicable. Preparation of subhalides by reaction of the elements is therefore always carried out using an excess of the respective halogen. In the field of tellurium subhalides not only the purity of the materials but also their quality and crystal size have been of considerable importance. The formation of single crystals of the subhalides has therefore been studied by annealing, Bridgman techniques and hydrothermal methods. Finally, it should be pointed out that tellurium subhalides may also be obtained by reactions using redox pairs such as TeX_4/SnX_2 (X = Br,I).

III.A Reaction of the Elements

Te_3Cl_2 is prepared by the reaction of dry chlorine with heated tellurium in a molar ratio of 2:1; the reaction is followed by homogenization, annealing and extraction of excess $TeCl_4$ [7,46]. The experimental set-up is described in Fig. 6. Powdered tellurium (~65 g) is placed into the horizontal tube and the joined piece is sealed off at A. In order to purify tellurium the horizontal tube is heated at 280 °C for several hours with simultaneous evacuation. The corresponding amount of chlorine is then condensed into the left leg during cooling with liquid nitrogen, and B is sealed. The chlorine gas is transported into the horizontal tube (200 °C) by gentle heating of the left leg. and cooling of the right leg of the arrangement. The left and the right legs are then sealed, the latter being first cleaned by careful sublimation of the reaction products into the horizontal part. The sealed tube containing the heterogeneous reaction products (Te, Te_3Cl_2, $TeCl_4$) is heated to yield a homogeneous

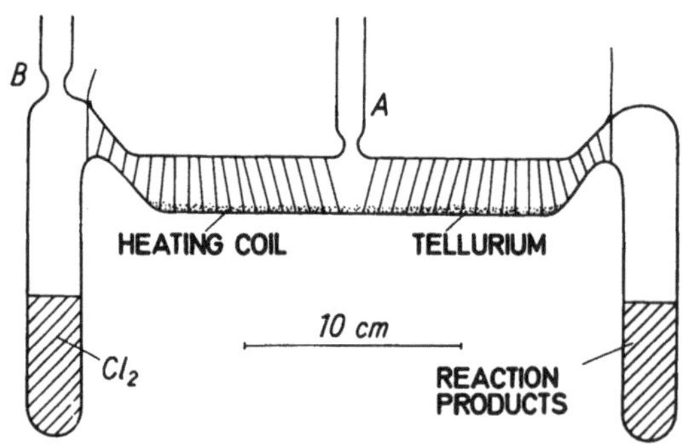

Fig. 6. Experimental arragement (quartz glass) for the preparation of Te_3Cl_2 by reaction of the elements [7]

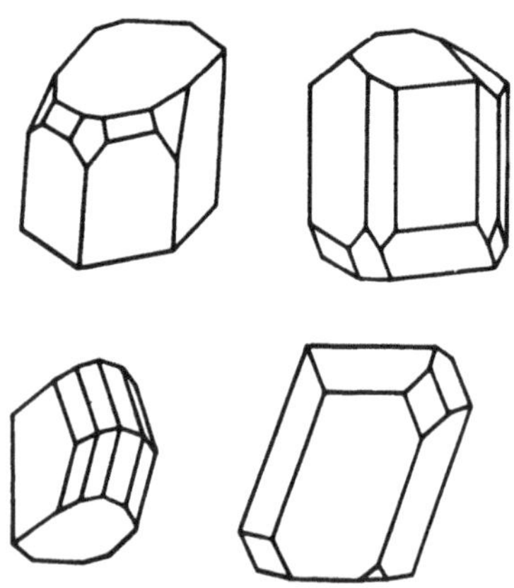

Fig. 7. Shapes of Te_3Cl_2 crystals obtained by annealing [44]

melt, quenched with ice water and annealed at 150 °C for two weeks. The final products consist of a mixture of compact black Te_3Cl_2 crystals with a high reflectivity and colourless-yellow crystals of $TeCl_4$. Crystals of Te_3Cl_2 grown by this method are up to two mm in size with a morphology of the crystals as shown in Fig. 7. A detailed description of the separation of mixtures of Te_3Cl_2 and $TeCl_4$ by extraction of $TeCl_4$ using a dry 1:1 solvent of toluene and light petroleum in an inert atmosphere is given in [7].

Te_2Cl is prepared [18] by use of the experimental arrangement and method for the reaction of the elements as given in Fig. 6. A 1:1 molar ratio of the elements chlorine and tellurium is allowed to react, and the horizontal tube is heated to 300 °C after the left and right leg of the construction have been sealed. Cooling from the melt at a controlled rate of 3 °C/min causes the formation of small Te_2Cl platelets on top of a solid bulk mixture. The condition of contact of the condensed phase with the corresponding vapour during cooling seems to be of general importance for the formation of the metastable crystalline phase Te_2Cl.

The preparation of *Te_2Br* [46] is carried out by means of a two-arm tube of quartz glass, one being filled with powdered tellurium and the other with liquid bromine; the molar ratio is 1:1. Tellurium is heated to 220–260 °C and bromine is warmed slowly from room temperature to 50 °C as the reaction proceeds. After consumption of the free halogen the reaction tube is heated to 300–350 °C and shaken. The homogeneous melt is quenched with ice water and the product annealed at 165 °C for two to three weeks in order to promote the establishment of the solid-state equilibrium. Dark needle-shaped crystals with a metallic lustre and up to 5 mm in length are obtained by this procedure as well as the corresponding amounts of the respective tetrabromide. Detailed informations on the $TeBr_4$ extraction from solid mixtures using diethyl phthalate at 80 °C as a solvent are given in [46].

α-TeI is prepared [66] by heating a mixture of tellurium and iodine (molar ratio 1:1) in a sealed ampoule of quartz glass to 200 °C for three hours. The reaction starts at 120 °C and is finished at about 200 °C. The ampoule is then cooled down to 172 °C and annealed at this temperature for a period of 17 days. After this time the surface of the ingot is covered with perfect compact α-TeI crystals which are nearly black and up to more than three mm in size.

III.B Hydrothermal Syntheses

The basic principles of hydrothermal synthesis performed in hydrohalic acids are described in [42,48]. With respect to tellurium subhalides hydrothermal syntheses with concentrated hydroiodic acid as a solvent represent successful methods for growing single crystals of subiodides Te_2I, β-TeI and α-TeI [11,18,41,44,49]. Since hydrohalic acids are very corrosive, use is made of quartz glass ampoules which can withstand hydrohalic acids even in the presence of free halogen and elements of group 6A. The experimental set-up used for hydrothermal syntheses in hydrohalic acids is schematically shown in Fig. 8.

In order to grow subiodide crystals the quartz glass ampoule is filled with powdered starting material (annealed mixtures of tellurium and iodine) and hydroiodic acid (67%) is added to 65% of the remaining free volume of the ampoule. The ampoule is cooled with liquid nitrogen, sealed off by simultaneous evacuation and placed

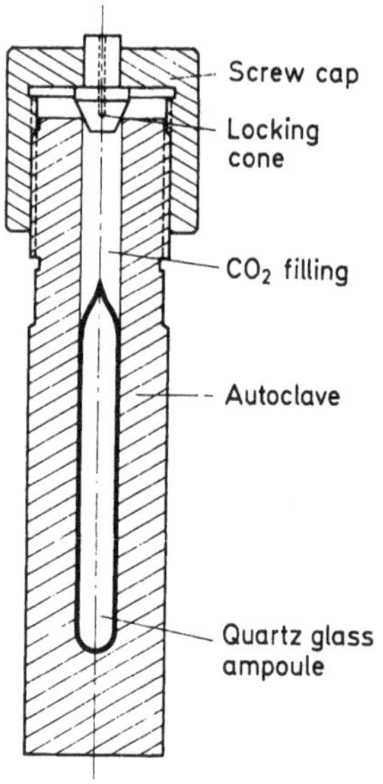

Fig. 8. Equipment for the hydrothermal method using hydrohalic acids as solvents [42,48].

into an autoclave. To prevent damage to the ampoule by the internal pressure developed during heating, the free volume of the autoclave is filled with a calculated amount of carbon dioxide. Table 2 contains detailed specifications of the experimental conditions with respect to hydrothermal growth of tellurium subiodides. Fig. 9 shows β-TeI crystals obtained by the hydrothermal method and Fig. 10 represents the intergrowth of α- and β-TeI crystals frequently observed at the bottom of the quartz glass ampoules.

It may be pointed out from the data given in Table 2 that stable α-TeI is the only phase which is formed under isothermal conditions. The metastable phases β-TeI

Table 2. Hydrothermal growth of tellurium subiodides in the set-up shown in Fig. 8 with the axis of the autoclave inclined ~45°.

	Te_2I	β-TeI	α-TeI
Nutrient[a]	$Te_{2.5}I$	$Te_{2.5}I$	$Te_{2.5}I$
Solvent	10 M HI	10 M HI	10 M HI
Degree of filling (%)	65	65	65
Temperature (°C)	265/280	192/198	195
Time (d)	10	7	8
Description	dark-metallic needles and plates, orthogonal, ≦10 mm, distributed over the whole ampoule; small amounts of β-TeI; nutrient Te and α-TeI	dark-metallic needles, sometimes with fibrous structure, partial sphe-rulites, ≦15 mm, distributed over the whole ampoule; nutrient Te and α-TeI	dark-metallic compact crystals, bladed, often twins, ≦3 mm, in direct contact with the nutrient (Te and α-TeI)

[a] Mixture of the elements, fused, then annealed at 150 °C for 3 h (at the bottom of the ampoule)

Fig. 9. β-TeI crystals grown by the hydrothermal method [18]

Fig. 10. Intergrowth of α-TeI (compact crystal) and β-TeI (fibrous crystal) [18]; SEM-photograph

and Te_2I are predominant when a temperature gradient is applied. The three iodides are usually prepared hydrothermally at temperatures above the peritectic reaction temperature of the system Te—TeI_4 [9] (see Fig. 3). It may be assumed that the solid subiodides are formed only after the cooling period of the autoclave has started.

The subhalides Te_2Cl, Te_3Cl_2 and Te_2Br have not been obtained by the method of hydrothermal syntheses [18,46].

III.C Bridgman Technique

The subhalides Te_3Cl_2, Te_2Br and α-TeI are stable intermediate phases (see phase diagrams in Figs. 1–3). Their incongruent melting behaviour is characterized by decomposition to solid tellurium + melt. On cooling homogeneous melts of subhalides the primary product of crystallization under conditions of thermodynamic equilibrium is therefore tellurium and not the respective subhalide. The situation is schematically shown in Fig. 11.

Using the Bridgman technique in order to grow large crystals of a peritectic intermediate phase (AB), the initial composition of the melt has to be chosen between

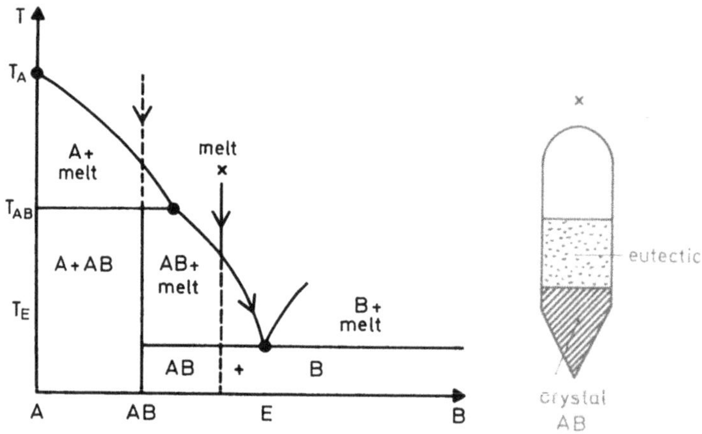

Fig. 11. Schematic representation of the conditions for growing crystals of a peritectic intermediate phase (AB) using the Bridgman technique

the peritectic and the eutectic composition (melt ×). In the course of crystal growth (AB) the composition of the melt changes in a direction which is indicated by the arrow on the liquides curve. Finally, the eutectic mixture (E) crystallizes on top of the crystal (AB).

In contrast to the equilibrium conditions in the Te—TeI_4 system given in Fig. 3, large *α-TeI* single crystals (Fig. 12) are obtained [50,67] by the Bridgman method using an initial melt with a 1:1 molar ratio of tellurium and iodine. The primary crystallization seems to be depressed by strong supercooling of the melt. This explanation agrees very well with the small temperature interval of about 15 °C in the two-phase region Te + liquid at a total composition TeI. In some events, even pure metastable *β-TeI* crystals (Fig. 13) have been obtained [50,68]. The appearance of these crystals may be connected with supercooling as well as partial metastable clustering within the corresponding melt (see also II.C).

Large *Te_2Br* crystals (Fig. 14) have been obtained [67] by the Bridgman technique using an initial melt of total composition ~40 atom-% bromine. The growth con-

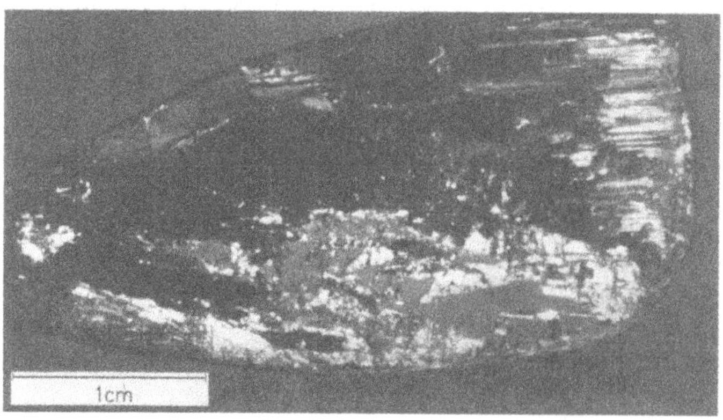

Fig. 12. α-TeI crystals grown by the Bridgman technique [67]

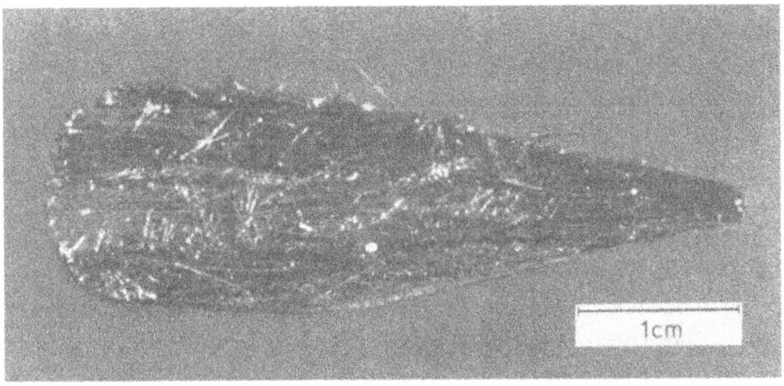

Fig. 13. β-TeI crystals grown by the Bridgman technique [68]

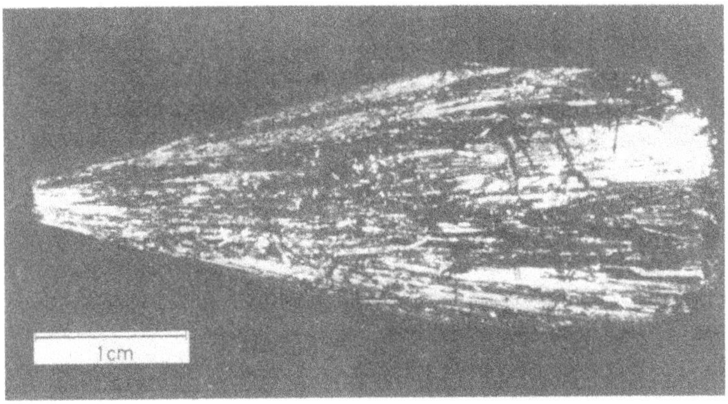

Fig. 14. Te_2Br crystals grown by the Bridgman technique [67]

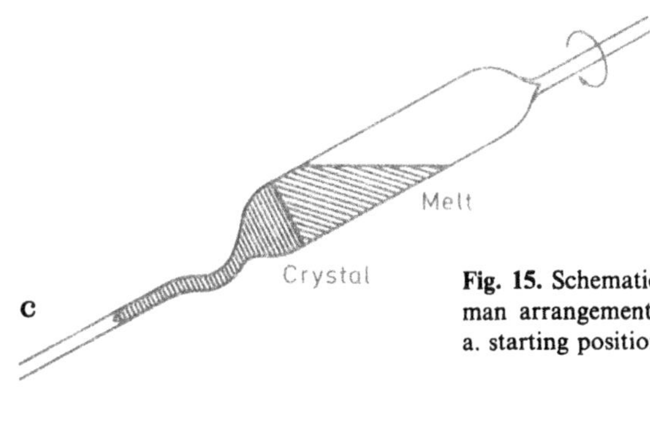

Fig. 15. Schematic representation of the modified Bridgman arrangement for the growth of Te_3Cl_2 crystals [69]. a. starting position; b. seeding; c. growth

ditions are in accordance with the schematic representation in Fig. 11 and with the phase diagram Te—$TeBr_4$ shown in Fig. 2.

Large crystals of *Te_3Cl_2* have been obtained [69] by a modified Bridgman method using an inclined ampoule (~30° to the horizontal level) which rotates around its central axis (0.3 rot/min). The experimental arrangement is schematically described in Fig. 15. The initial composition of the melt was 56 atom-% chlorine and 44 atom-% tellurium which is completely molten at the temperatur of the peritectic reaction isotherm (see Fig. 1). The shape of the ampoule allows simple seeding before the start of the growth process by turning the ampoule from position a to b; the lower part of the seed is held at a temperature <200 °C. Rotation of the ampoule after inclination (position c) leads to continuous stirring of the melt which results in lower supercooling and a higher temperature gradient at the interface. The furnace is moved at a speed of 7.5 mm/day. A Te_3Cl_2 crystal obtained by this modified Bridgman technique is shown in Fig. 16.

Fig. 16. Te_3Cl_2 crystals grown by the modified Bridgman technique as described in Fig. 15

III.D Redox Systems TeX_4/SnX_2 (X = Cl, Br, I)

Reactions of SnX_2 with TeX_4 have been investigated by DTA and X-ray analyses [30,70], Sn(II) being always oxidized to Sn(IV); simultaneous reduction of the respective tellurium tetrahalide takes place as follows:

$$2\,SnCl_2 + TeCl_4 \rightarrow Te + 2\,SnCl_4 \qquad (6)$$

$$7\,SnBr_2 + 4\,TeBr_4 \rightarrow 2\,Te_2Br + 7\,SnBr_4 \qquad (7)$$

$$3\,SnI_2 + 2\,TeI_4 \rightarrow 2\alpha/\beta\text{-}TeI + 3\,SnI_4 \qquad (8)$$

No formation of $TeCl_2$ [24,25] (see also I.B) in the chloride system has been observed. The formation of α- or β-TeI depends on the maximum temperature of the reaction. Fig. 17 shows the reaction of SnI_2 with TeI_4 monitored by DTA. The broad ΔT-peak occurring between room temperature and 100 °C corresponds to the reaction of the initial components. The reaction products, SnI_4 and tellurium monoiodide, melt

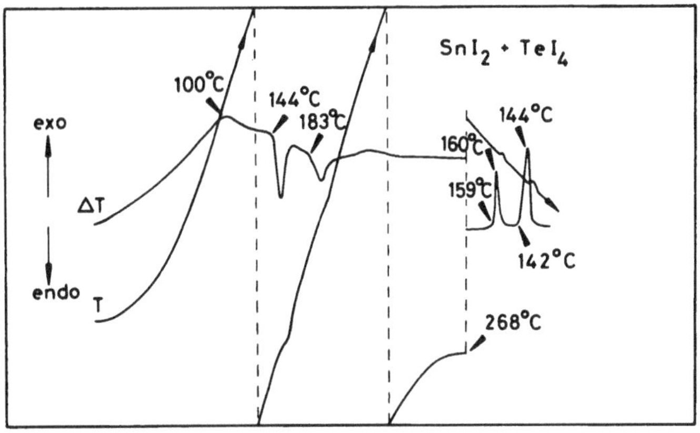

Fig. 17. Reaction of SnI_2 with TeI_4, monitored by DTA [30]

at 144 and 183 °C, respectively. During cooling from the melt supercooled solidification of TeI is followed by only slightly supercooled crystallization of SnI_4. The crystalline subiodide formed by cooling from the melt was shown to be β-TeI. On the other hand, when annealing a mixture of SnI_2 and TeI_4 at 100 °C for two weeks, α-TeI is produced as the only subiodide.

With regard to reactions (7) and (8) an excess of tellurium tetrahalide results in the excess remaining in the reaction product. If an excess of Sn(II) is used, further reduction of the subhalide to elemental tellurium is observed:

$$SnBr_2 + 2\,Te_2Br \rightarrow 4\,Te + SnBr_4 \qquad (9)$$

$$SnI_2 + 2\alpha\text{-}/\beta\text{-}TeI \rightarrow 2\,Te + SnI_4 \qquad (10)$$

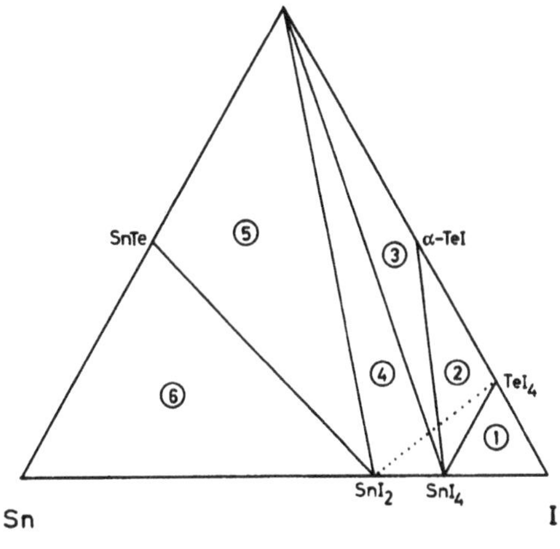

Fig. 18. Quasibinary cross-sections in the ternary system Sn—Te—I [70]. Three-phase equilibria: 1. SnI_4 + TeI_4 + I; 2. SnI_4 + α-TeI + TeI_4; 3. SnI_4 + Te + α-TeI; 4. SnI_2 + Te + SnI_4; 5. SnTe + Te + SnI_2; 6. Sn + SnTe + SnI_2

The dependence of the degree of reduction of Te(IV) on the amount of Sn(II) involved in the reactions is quite easily understood by taking into account the phase relations in the respective ternary systems Sn—Te—X (X = Cl, Br, I). In the case of the iodide system solid-state phase equilibria have been investigated [70]. As can be seen from Fig. 18 there are five quasibinary cross sections dividing the ternary system into six areas of three-phase equilibria. The dotted line represents a pseudo-binary section including the reactions between SnI_2 and TeI_4. Depending on the molar ratio of the redox components, five different phase equilibria are observed: SnI_4 + TeI_4 + α-TeI; 2 α-TeI + 3 SnI_4; SnI_4 + Te + α-TeI; 2 SnI_4 + Te and SnI_2 + SnI_4 + Te. In this way, the reduction of Te(IV) to Te(I) or elemental tellurium is quantitatively described by the phase relations of the respective ternary system.

IV Crystal Structures

IV.A Crystallographic Data

Single crystals of tellurium subhalides have been investigated by X-ray methods; Te_3Cl_2 [18,44,45,46,49], Te_2Cl [18,44,45,49], Te_2Br [18,44,45,46,49], Te_2I [18,44,45,49], β-TeI[18,41,44,45,47,49], and α-TeI [17,18,44,45,47,49,66]. The final lattice parameter values based on photographic investigations as well as the refinement of the cell parameters using an automated X-ray diffractometer are given in Table 3. Space groups are confirmed by complete crystal structure analyses.

The monoiodide (Te_nI_n) at first was reported [17] to be orthorhombic with possible space groups Cmma or Cm2a, Z = 16 Te_2I_2, D_x = 5.40 g · cm^{-3} and cell parameters a = 823, b = 3000, c = 997 pm. These lattice constants can be fairly well

Table 3. Crystallographic data for tellurium subhalides [44].

Compound	Symmetry Space Group	a α	b β	c (pm) γ (°)	Z	D_x D_0 (g · cm^{-3})
Te_3Cl_2	mcl $P2_1/n$	1013.6 (15)	863.5 (3) 100.74 (6)	703.9 (8)	4	4.98 4.90
Te_2Cl	orh Pnam	1482	1281	400	8	5.08
Te_2Br	orh Pnam	1492.1(12)	1284.3 (6)	400.5 (3)	8	5.80 5.61
Te_2I	orh Pnam	1529.5 (3)	1369.4 (4)	412.3 (2)		5.88 5.86
β-TeI	mcl C2/m	1538.2 (2)	418.2 (2) 128.09 (1)	1199.2 (2)	8	5,56 5.57
α-TeI	trcl $P\bar{1}$	995.8 (3) 104.37 (1)	799.2 (20) 90.13 (4)	821.2 (4) 102.89 (7)	8	5.49 5.41

reproduced [66] using the cell parameters of the triclinic cell given in Table 3 and the conversion matrix:

$$\begin{matrix} 0 & 0 & 1 \\ 1 & 4 & 1 \\ 1 & 0 & 0 \end{matrix}$$

Furthermore, it was shown [66] that Weissenberg photographs of twinned α-TeI crystals give the suggestion of the Laue group mmm. Thus, it appears that Te_nI_n [17] was erroneously determined as orthorhombic and is the the triclinic phase α-TeI instead.

The subhalides Te_2Cl, Te_2Br and Te_2I are isotypic compounds. This has been of special importance with respect to the X-ray crystal structure analyses of the subiodides. Tellurium and iodine positions in Te_2I were distinguished in accordance with the tellurium and bromine positions in Te_2Br. These results were then used for the interpretation of the β- and α-TeI crystal structures.

IV.B Structural Units

The crystal structures of tellurium subhalides are built up of structural units presented in Fig. 19. Because of the high tellurium content of these compounds, similarities with the latter and the structural properties of elemental tellurium [53] could be expected and, in fact, were confirmed [18,44,45,47].

The structural unit of Te_3Cl_2 is an unlimited chain of tellurium atoms twisted around a crystallographic 2_1 axis. This arrangement is similar to that in elemental tellurium where the generating screw is of 3_1 or 3_2 symmetry, respectively. Every third tellurium atom in Te_3Cl_2 is bound to two chlorine atoms in axial positions which

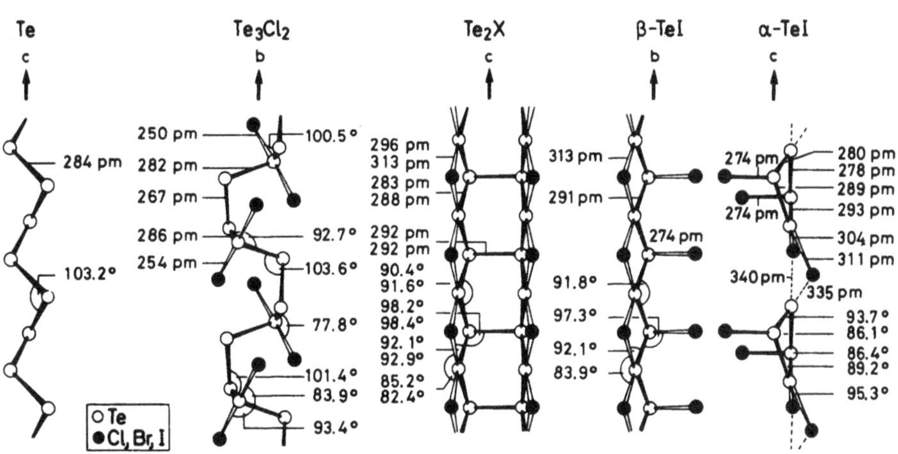

Fig. 19. Structural units of tellurium and tellurium subhalides [44]: "modified tellurium structures". The upper numbers (bond lengths and bond angles) in Te_2X refer to X = Br, the numbers below to X = I

account for the observed composition. The macromolecular structural units of the isotypic subhalides with the general composition Te_2X may be described as chains of fused six-membered tellurium rings with a boat conformation, and with bridging halogen atoms at their borders. The β-TeI structural unit consists of a planar tellurium zigzag chain with bridging and terminal iodine atoms alternating along the chain. This arrangement is similar to that in Te_2X with the terminal iodine of the β-TeI unit being replaced by a Te—Te bond between two neighbouring chains. α-TeI does not contain macromolecular structural units; the $1/\infty$ homonuclear chains and screws of tellurium atoms are degenerate in the α-TeI structure yielding a four-membered tellurium ring, with the four tellurium atoms showing three different arrangements of coordination. The molecules of composition Te_4I_4 are linked by short intermolecular Te ... I contacts along the crystallographic c-axis direction; the conformation of the Te_4-ring [44] is shown in Fig. 20.

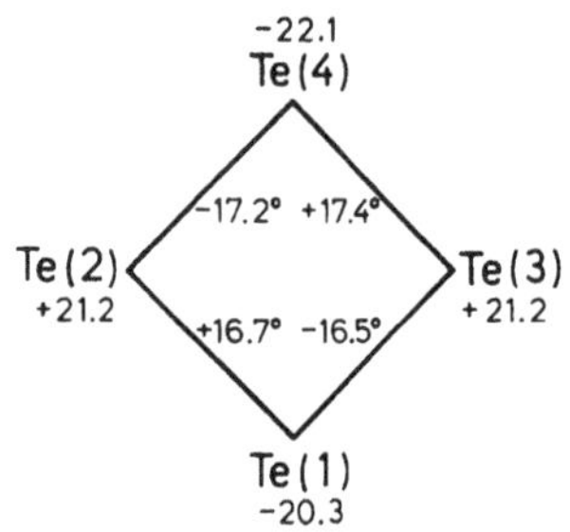

Fig. 20. Conformation of the Te_4-ring in α-TeI [44]. Torsion angles and atomic distances (pm) from a plane fitted to the ring (schematically)

With respect to the homonuclear linkage of tellurium atoms, the tellurium subhalides may be described as "modified tellurium structures". The 3_1-chain of the element is modified by steps, going from a 2_1-chain in Te_3Cl_2 to a ribbon of condensed planar zigzag chains in Te_2X and a separated planar zigzag chain in β-TeI down to a four-membered tellurium ring in α-TeI.

Coordination geometries of tellurium in tellurium subhalides clearly follow from a consideration of the structural units presented in Fig. 19. Even if the conditions of chemical bonding in these units are not quite clear at first sight, they are easily integrated into accepted concepts using the common valence rules in connection with Gillespie-Nyholm rules [71] and taking into account the different coordination

Table 4. Formal states of oxidation and proposed configurations of tellurium atoms within the crystal structures of tellurium subhalides.

Compound	Modified Formula	Te: State of Oxidation and Configuration
Te	$1/\infty\{Te^0\}$	Te^0 (CN=2)
Te_3Cl_2	$1/\infty\{Te^0_2[TeCl_2]\}$	Te^0 (CN = 2) + Te^{2+} (ψ-trigonal bipyramidal)
Te_2X	$1/\infty\{Te^+[TeX]^-\}$	Te^+ (ψ-tetrahedral) + Te^0 (ψ-octahedral)
β-TeI	$1/\infty\{[TeI]^+ [TeI]^-\}$	Te^{2+} (ψ-tetrahedral) + Te^0 (ψ-octahedral)
α-TeI	cycl $\{Te^0[TeI]_2^+ [TeI_2]^{2-}\}$	Te^0 (CN=2) + Te^{2+} (ψ-tetrahedral) + Te^0 (ψ-octahedral)

numbers of tellurium [72]. Formal oxidation states and proposed configurations of tellurium atoms in the subhalide structures, as well as modified formulas taking into account the different coordination behaviour of tellurium atoms, are given in Table 4.

Finally, it may be pointed out that the homonuclear arrangement of tellurium atoms in the Te_2X structural units is equal to that of germanium atoms in Sr_3Ge_4 [73]. This corresponds to the concept that isoelectronic species, such as $1/\infty\ Ge_2^{3-}$ and $1/\infty\ Te_2^{1+}$, show close relationships concerning their chemical and structural behaviour. Numerous examples of this type are discussed in [72].

IV.C Structural Packing

Projections on the crystal structures of tellurium subhalides are shown in Fig. 21. Their three-dimensional structural arrangements [44] are of a similar heterodesmic type which is also present in the crystal structure of elemental tellurium [53]. The shortest intermolecular distances in subhalide structures are in the range of 75–80% of the corresponding sums of the van der Waals radii [74]. In the structure of the element, the two-fold coordination of tellurium atoms within the chains is completed to a 2+4-octahedron by intermolecular contacts to neighbouring chains. The consideration of an increase in the coordination number of tellurium in the subhalide structures by considering short intermolecular contacts works only in the case of Te_3Cl_2: the Ψ-trigonal bipyramidal configuration around the tellurium atom completes its coordination sphere to a 4+2-octahedron by contacts (336.3 and 340.0 pm) to chlorine atoms of a neighbouring screw. The resulting geometry is presented in Fig. 22.

The observed cleavage of subhalide crystals agrees very well with the special packing of the molecular units in the corresponding crystal structures: Te_3Cl_2 shows a perfect cleavage parallel to $(10\bar{1})$; there are two nearly equivalent planes of cleavage ((100) and $(20\bar{1})$) in the β-TeI structure which account for the fibrous mechanical behaviour. The homogeneous surrounding of the molecular Te_2X-units makes the material brittle. α-TeI undergoes perfect cleavage parallel to (100) and (010).

V Mössbauer Investigations

Mössbauer spectroscopy has been used to study a number of halogen-containing compounds of tellurium in which the formal oxidation state of tellurium is II [26] (see also I.B), IV [75,76] or VI [75,76]. The knowledge of the existence of tellurium subhalides with low formal oxidation states of the chalcogen gave rise to Te-125 Mössbauer investigations [78] of the subhalides, and additionally to I-129 Mössbauer studies on α-TeI [75].

The Te-125 Mössbauer spectra [78], at 4.2 K, of polycrystalline tellurium subhalides Te_3Cl_2, Te_2X (X = Br, I), β-TeI and α-TeI show the presence of at least two

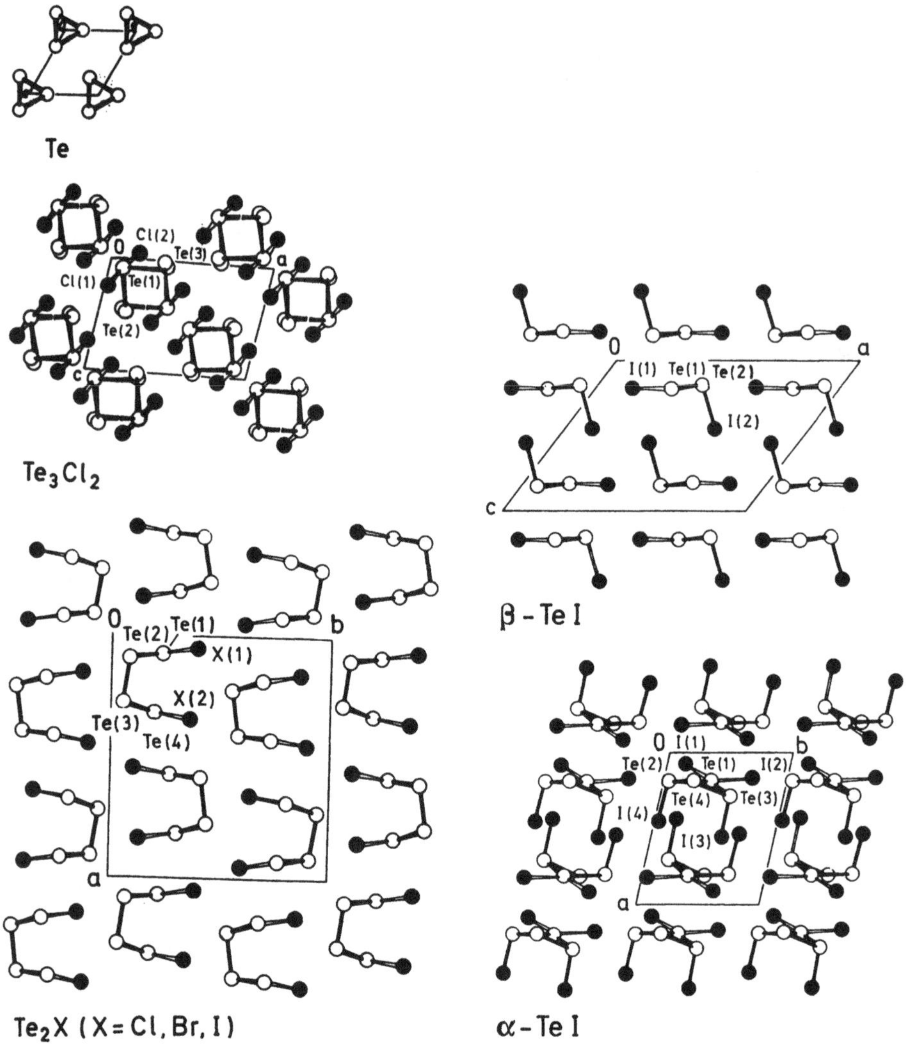

Fig. 21. Arrangements of the structural units (Fig. 19) of tellurium and tellurium subhalides in the corresponding crystal structures [44]. Projections along the macromolecular structural units; α-TeI is projected along c*

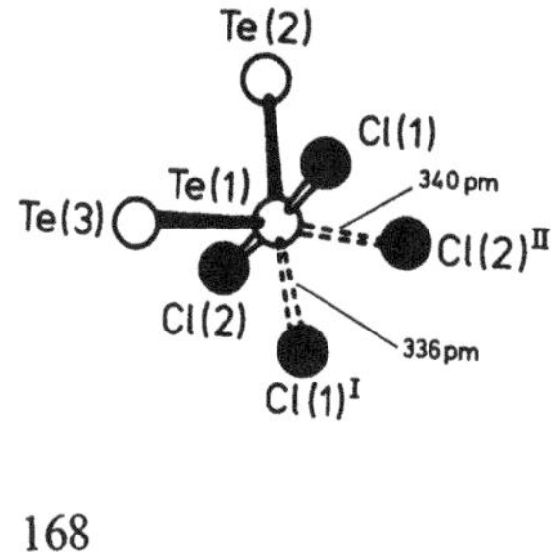

Fig. 22. Te(1)-coordination (4 + 2-octahedron) in Te_3Cl_2, generated by taking into account the short intermolecular contacts [44].

Te(2) — Te(1) ... $Cl(2)^{II}$	88.11(9)°
$Cl(2)^{II}$ ... Te(1) ... $Cl(1)^{I}$	104.94(12)
$Cl(1)^{I}$ ... Te(1) — Te(3)	84.97(11)
Cl(1) — Te(1) ... $Cl(1)^{I}$	73.92(14)
Cl(1) — Te(1) ... $Cl(2)^{II}$	108.48(15)
Cl(2) — Te(1) ... $Cl(1)^{I}$	108.42(14)
Cl(2) — Te(1) ... $Cl(2)^{II}$	73.19(14)
Te(2) — Te(1) ... $Cl(1)^{I}$	156.94(8)
Te(3) — Te(1) ... $Cl(2)^{II}$	150.92(9)

different tellurium sites in each subhalide which is in general in accordance with the results of crystal structure analyses (IV.B). The chemical isomer shifts δ (relative to $Zn^{125m}Te$) in four different types of coordination have been found to be 0.6 mm s^{-1} for trigonal bipyramidal tellurium atoms ($[Te(Te)_2(Cl)_2E]$ in Te_3Cl_2), 0.8–1.0 mm s^{-1} for square planar (pseudooctahedral) tellurium atoms ($[Te(Te)_2(X)_2E_2]$ with X = Br, I in Te_2Br, Te_2I, β-TeI and α-TeI), 0.3–0.8 mm s^{-1} for trigonal pyramidal (pseudotetrahedral) tellurium atoms ($[Te(Te)_2XE]$ with X = Te or I in Te_2Br, Te_2I, β-TeI and α-TeI) and 0.8 mm s^{-1} for two-coordinate (pseudotetrahedral) tellurium atoms ($[Te(Te)_2E_2]$ in Te_3Cl_2). The corresponding quadrupole splittings for the four environments were 11.0, 12.1–13.7, 3.4–6.1 and 8.1 mm s^{-1}, respectively. No distinction was made between the square planar and the formally two-coordinate tellurium in the Te_4I_4-ring of α-TeI (see Figs. 19 and 20).

The I-129 Mössbauer spectrum [75] of polycrystalline α-TeI at 4.2 K shows the presence of two inequivalent iodine atoms. The iodine atom with large isomer shift δ (0.84 mm s^{-1} relative to ZnTe) was identified as the terminal iodine which is bound to a trigonal pyramidal coordinated tellurium; the iodine atom with a small isomer shift δ (0.11 mm s^{-1} relative to ZnTe) was identified as a bridging iodine between two Te_4I_4 molecules which are neighbours in the crystallographic c-axis direction. A small coupling constant of the latter iodine atom reflects the relatively weak I-Te contact as compared with the strong bond between tellurium and the terminal iodine. This agrees very well with the result of the crystal structure analysis (see Figure 19) revealing bond lengths Te (trig. pyr.)-I = 274 pm and Te (squ. plan)-I = 304–311 pm.

VI Ternary Subhalides

Tellurium subhalides contain the series of isotypes Te_2X (X = Cl, Br, I). Solid solutions of the stable Te_2Br and the metastable Te_2I have been studied. Further examples of studies of ternary subhalides were based on selenium subhalides, with SeCl [79] and α-/β-SeBr [80,81] being the only known subhalides; no intermediate binary compound exists in the Se-I system [82]. The latter fact in particular stimulated the investigation of the ternary Se-Te-I system in order to find out whether there are distinct intermediate ternary subiodides and/or regions of ternary solid solutions based on the subiodide α-TeI.

VI.A $Te_2Br_{1-x}I_x$

VI.A.1 Vitreous Phase

Homogeneous glasses of composition $Te_{2.0}Br_{1-x}I_x$ [54] with a high refractive index and a silver-grey metallic lustre are obtained by cooling the melts ($0 \leqq x \leqq 0.75$) from 300 °C to room temperature and by quenching with liquid nitrogen ($x > 0.75$), respectively. In the course of heating the vitreous phases an exothermal transformation to a crystalline state takes place and this transition is followed by endothermal melting. A glass of composition $Te_{2.0}Br_{0.75}I_{0.25}$, for example, is first transformed to the corresponding crystalline state at 138 °C and then melts in the range of 192–212 °C. A different behaviour is shown by glasses of composition $x > 0.75$. Here, the exothermal transformation starts at 30 °C and does not result in mixed crystals but in a combination of Te, Te_2Br, β-TeI and amorphous phases. Glassy $Te_{2.0}Br_{0.75}I_{0.25}$ was investigated by X-ray diffraction [54]. The raw scattered X-ray intensity curve is shown in Fig. 23; the radial distribution function (RDF) deduced from these data is given in Fig. 24. The nearest neighbour distribution peak at 280 pm is slightly shorter than the nearest neighbour position of 291 pm expected for the corresponding molecular unit which is shown in Fig. 19. The second neighbour distribution with a peak at 410 pm is consistent with an average bonding angle of 91°, similar to that of the crystalline state. Except for the weak peaks at 620 and 790 pm the RDF converges rapidly to the uniform atomic density.

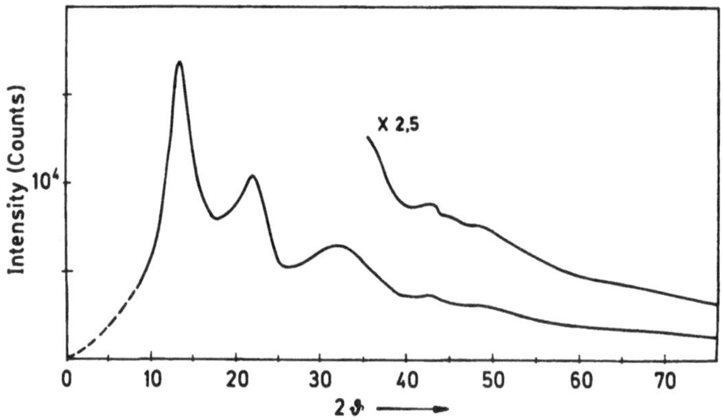

Fig. 23. Scattered X-ray intensity of glassy $Te_{2.0}Br_{0.75}I_{0.25}$ obtained by MoKα-radiation [54]. The dashed line denotes the quadratic extrapolation to zero-scattering angle

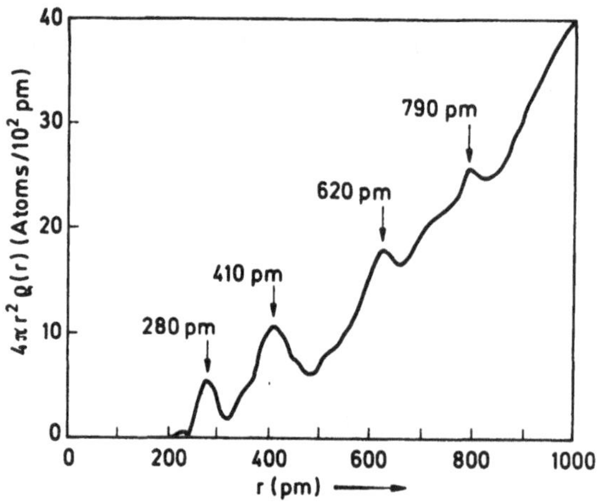

Fig. 24. Radial distribution function of glassy $Te_{2.0}Br_{0.75}I_{0.25}$ obtained by Fourier inversion of the data in Fig. 23. The peak positions are indicated by arrows

The tellurium substructure of crystalline Te_2X consists of two parallel zig-zag chains that are cross-linked at every second atom along the chain. Using an average neighbour separation of 280 pm and an average bond angle of 91°, the Te—Te third and fourth nearest neighbours within the free chains lie at 640 and 820 pm, respectively. Thus, the weak peaks at 620 and 790 pm indicate that the chain structure of the crystalline state is maintained over a distance of ~1600 pm in the glassy state. The long-range order is probably lost because of a breakdown of correlation between the chains; twisting and bending of the chains would also destroy some of the correlations between the atoms in the same chain.

This model, in general, is in accordance with the Mössbauer spectrum of the glassy phase [78] which is different from those of the corresponding crystalline compounds, but again shows the existence of at least two different tellurium sites.

VI.A.2 Crystalline Phase

Mixed crystals $Te_{2,0}Br_{1-x}I_x$ with $x < 0.75$ are easily prepared by annealing the respective vitreous form which can be recovered from the melt in a reversible process. The reversibility indicates the close relationship between the glassy and the crystalline state. Crystallized samples were investigated by the X-ray powder technique and were shown to consist of single-phase solid solutions [18]. The cell volume versus composition plot is given in Fig. 25 and obeys to Vegard's rule.

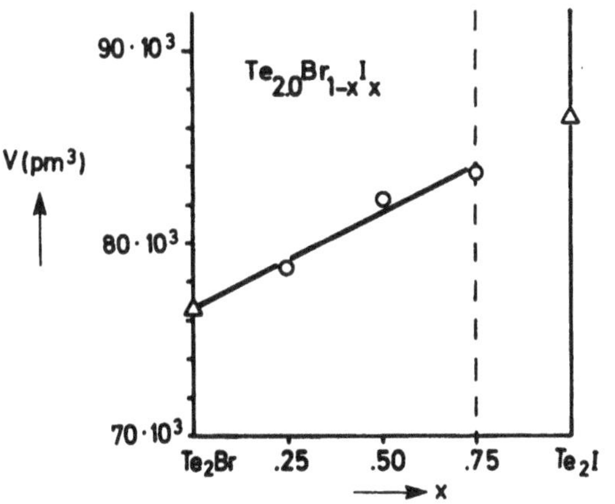

Fig. 25. Cell-volume vs. composition plot of mixed crystals $Te_{2.0}Br_{1-x}I_x$ [18]

VI.A.3 Photoelectric Properties

The density of valence states of crystalline $Te_{2.00}Br_{0.75}I_{0.25}$ was determined [54] by UV and X-ray photoemission experiments and compared with the density of valence states of elemental tellurium [83]. The results are given in Figs. 26a–c.

The Te 4d level in the compound has shifted by only 1.0 eV to a higher energy (less bound) with respect to its position in the elemental form. The differences in the valence bands of the compound and the element are clearly observed. The sharp peak at 2 eV in Te is attributed to the upper nonbonding lone electron pairs associated with twofold coordinated tellurium. In the mixed crystal this peak is absent, suggesting that the lone pair structure typical of the element is lost. The

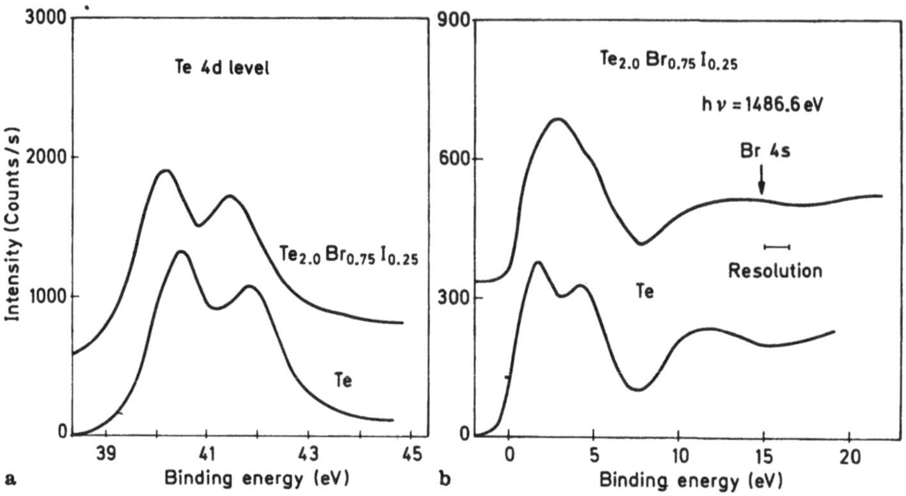

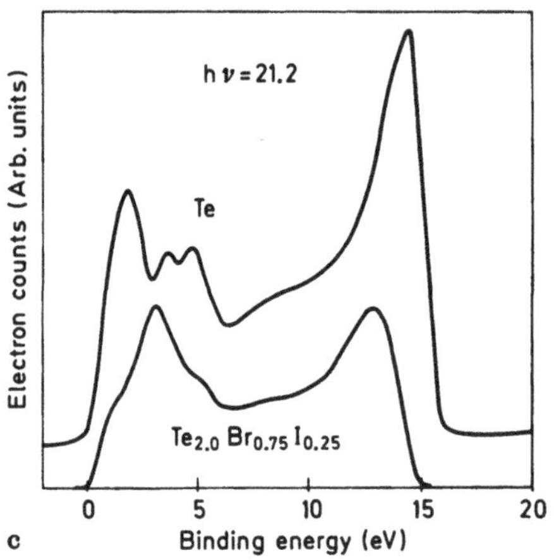

Fig. 26. Photoelectric properties of mixed crystals $Te_{2.0}Br_{0.75}I_{0.25}$ [54].
a. Comparison of the Te 4d level in $Te_{2.0}Br_{0.75}I_{0.25}$ and elemental Te (from Ref. [83]).
b. Comparison of the X-ray (hν = 1486.6 eV) photoemission from the valence bands of $Te_{2.0}Br_{0.75}I_{0.25}$ and elemental Te (from Ref. [83]).
c. Comparison of the UV (hν = 21.2 eV) photoemission from the valence bands of $Te_{2.0}Br_{0.75}I_{0.25}$ and elemental Te (from Ref. [83])

peak near 12 eV in the valence band of the compound is broader in the compound than in elemental tellurium, most likely because of the contribution from the Br and I s-levels which lie in the region of 13–15 eV.

VI.B α-$Te_{1-x}Se_xI$

The ternary system Se—Te—I was investigated by differential thermal analysis and X-ray powder diffraction [84]. The resulting solid-state phase equilibria are shown in Fig. 27. No distinct (new) ternary phase appears in the system but there is a region of intermediate solid solution α-$Te_{1-x}Se_xI$ with $x \leqq 0.18$ (B in Fig. 27).

The chemical behaviour of mixed crystals, Se_xTe_{1-x}, in the ternary systems does not change continuously but shows an abrupt break: solid solutions with $x < 0.44$ behave like tellurium and those with $x > 0.44$ show a behaviour similar to that of elemental selenium. From the point of view of chemical reactions it is interesting to note that selenium reacts with TeI to give TeI_4 and (depending on the Se-to-Te ratio applied) mixed crystals of composition Se_xTe_{1-x} ($x \geqq 0.44$).

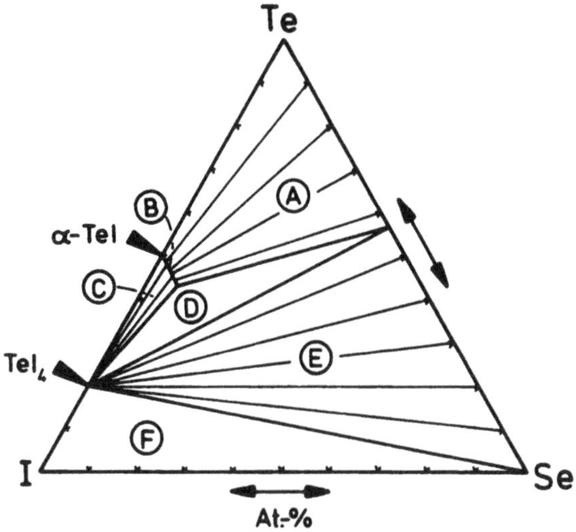

Fig. 27. Solid-state phase equilibria in the ternary Se—Te—I system [84].
A: two-phase domain:
α-$Te_{1-x}Se_xI$ ($x \leqq 0.18$)
+ Te_xSe_{1-x} ($x \geqq 0.56$)
B: one-phase domain:
α-$Te_{1-x}Se_xI$ ($x \leqq 0.18$)
C: two-phase domain:
α-$Te_{1-x}Se_xI$ ($x \leqq 0.18$) + TeI_4
D: three-phase domain:
TeI_4 + α-$Te_{.82}Se_{.18}I$ + $Te_{.56}Se_{.44}$
E: two-phase domain:
TeI_4 + Se_xTe_{1-x} ($x \geqq 0.44$)
F: three-phase domain:
I + TeI_4 + Se

VII Optical Properties

The first optical investigations of tellurium subhalides have been the determination of the spectral reflectances of powdered samples of the stable subhalides Te_3Cl_2, Te_2Br and α-TeI relative to MgO as a standard [7,9]. The increase of the logarithm of the reciprocal relative reflectance plotted against the photon energy was correlated with the respective band gap: Te_3Cl_2 1.3 eV, Te_2Br 0.6 eV, α-TeI 1.0 eV. Raman and infrared spectroscopy were then used to investigate small crystals of α-TeI at room temperature [66]. It was found that some of the zone-center optical phonons are Raman- as well as infrared-active, and it was concluded that α-TeI belongs to a non-centrosymmetric space group. Because of its optical properties, α-TeI was expected to be suitable for nonlinear optical devices. After the successful growth of large single crystals of tellurium subhalides (seee III.C) reinvestigations of the optical properties of α-TeI [85,86,87,88] showed no Raman-active modes which were infrared-active. This confirmed the result of the crystal structure analysis, i.e. centrosymmetric space group $P\bar{1}$ for α-TeI (see IV).

Detailed optical investigations of tellurium subhalides have been carried out on Te_3Cl_2 and α-TeI which, from the point of view of, "modified tellurium structures" are the structural antipodes.

VII.A Absorption

Figure 28 shows the absorption spectra of single crystals of Te_3Cl_2 and α-TeI [50,85,86]. The spectral dependence of the absorption constant may be interpreted as being characteristic of an allowed transition. These results show the (direct) gap energy at 300 K to be E_{gap} = 1.52 and 1.32 eV, respectively which is markedly larger than

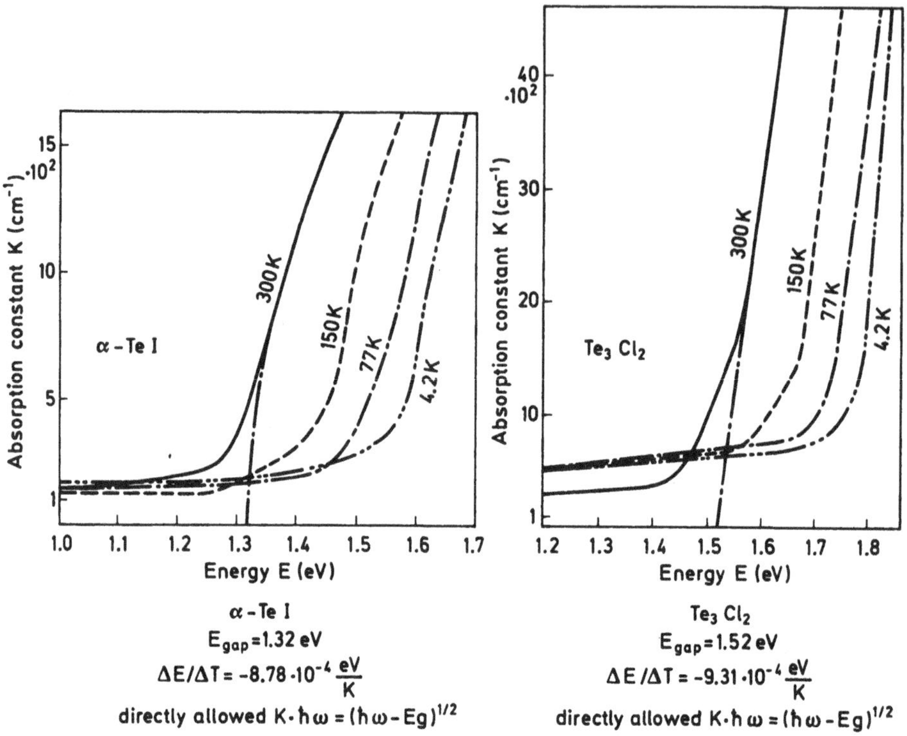

Fig. 28. Optical absorption spectra of α-TeI and Te_3Cl_2 [85,86]

that of Te with 0.34 eV [85]. The temperature dependence $\Delta E/\Delta T$ is $-9.31 \cdot 10^{-4}$ eV/K for Te_3Cl_2 and $-8.78 \cdot 10^{-4}$ eV/K for α-TeI [50]. The dielectric constants of the subhalides show a characteristic increase from Te_3Cl_2 ($\varepsilon_\infty = 6$) to α-TeI ($\varepsilon_\infty = 12$) and to Te ($\varepsilon_{\infty\,\parallel} = 23$; $\varepsilon_{\infty\,\perp} = 36$) reflecting the increase of the polarizability.

VII.B IR/Raman

The contribution of the lattice vibrations to the optical constants of Te_3Cl_2 and α-TeI was determined by measuring the reststrahlen spectra in the FIR. The frequency dependence of the real and imaginary part of the DK was calculated by performing a Kramers-Kronig analysis [85,86,87]. Figure 29 shows the reflectivity spectra of Te_3Cl_2 for the polarization directions $E \parallel b$ and $E \perp b$ and for α-TeI in the polarization directions $E \perp c$ and $E \parallel c$ at 20 K. The reststrahlen spectrum of Te_3Cl_2 is characterized by strongly polarization-dependent structures. In both polarization directions four strong phonon peaks can be resolved. The bands near 100 cm^{-1} can be attributed to Te-chain vibrations while it is suggested that the high-energy terms are Cl-vibrations. The variety exhibited by the reststrahlen

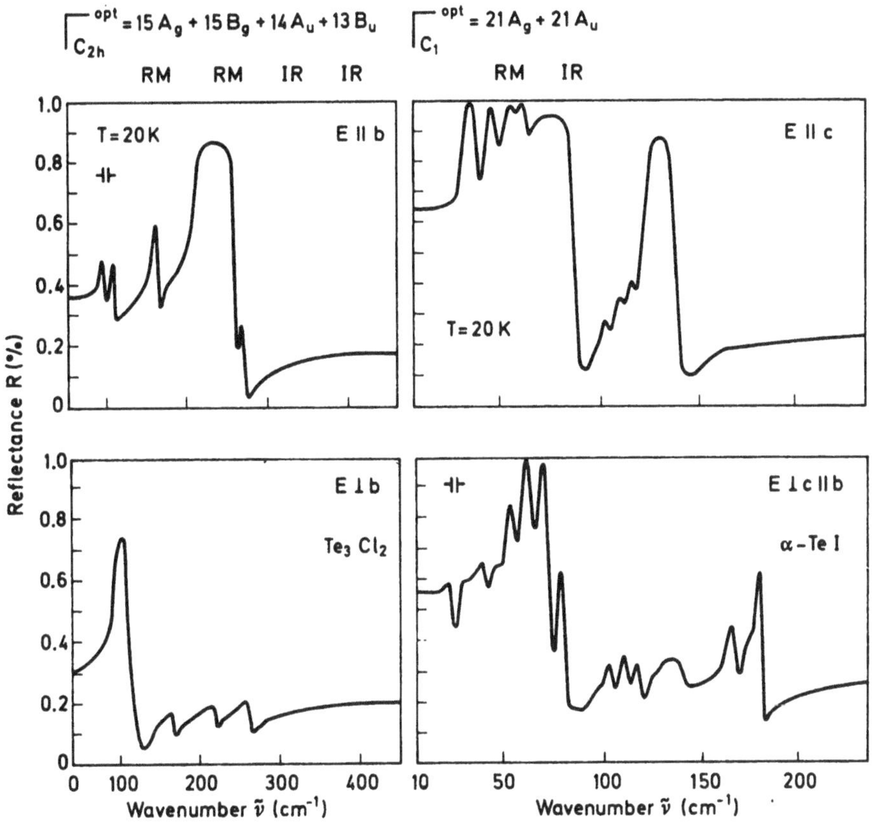

Fig. 29. Reflectivity spectra of Te_3Cl_3 and α-TeI [85, 86]

spectrum of α-TeI is characteristic of a molecular crystal. An interesting fact is the polarization independence of the structure at $100\ cm^{-1}$, indicating isotropic Te-ring breathing modes.

Figure 30 shows the contribution of the lattice vibrations to the optical constants of Te_3Cl_2 and α-TeI for both polarization directions. Two dominant oscillators appear at $224\ cm^{-1}$ $E \parallel b$ and $108\ cm^{-1}$ $E \perp b$ in Te_3Cl_2. The large Lo-To split in Te_3Cl_2 is caused by the static electric charge induced by the polarization of the Te-atoms by the electronegative Cl-atoms. In α-TeI also two dominating oscillators are observed which are not characterized at present.

Raman spectra of α-TeI and Te_3Cl_2 [85, 86] (Fig. 31) confirm the alternative prohibition of infrared-active and Raman-active modes. It should be mentioned in this context that resonant Raman scattering experiments on Te_3Cl_2 at He-temperatures seem to be possible.

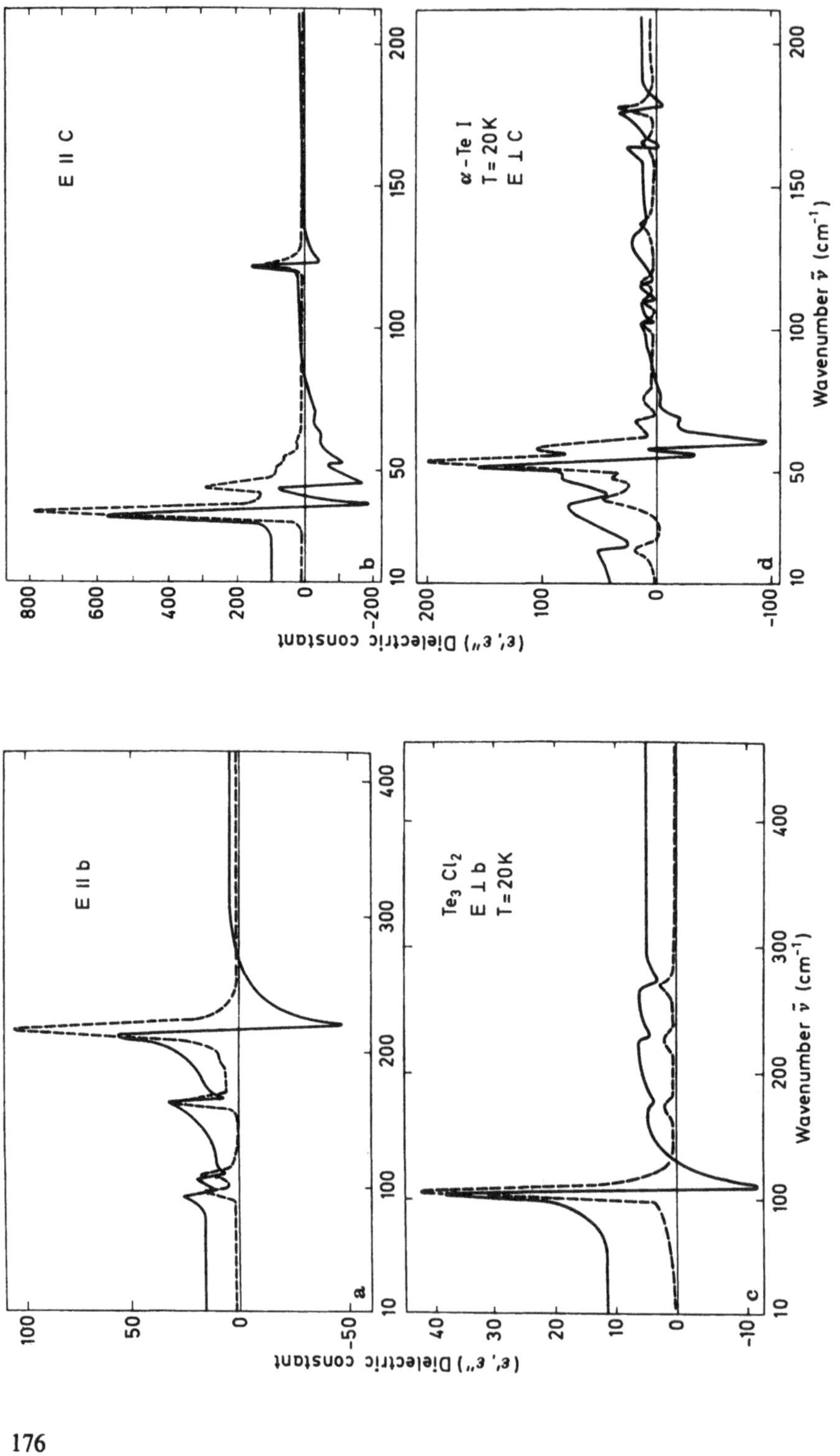

Fig. 30. Kramers-Kronig spectra of Te_3Cl_2 and α-TeI [85,86]

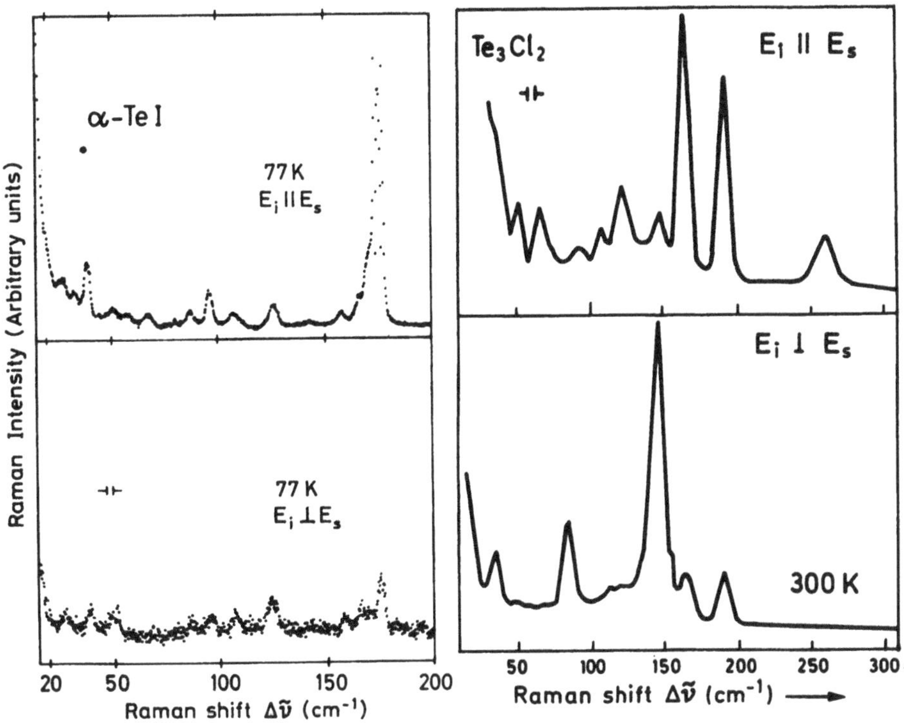

Fig. 31. Raman spectra of α-TeI and Te_3Cl_2 [85,86]

VIII Electrical Conductivity

The first subhalide which was investigated by measurement of the electrical conductivity was tellurium monoiodide with a 1.1 eV band gap calculated from the temperature dependence of the electrical conductivity [89]. After the crystal structure determination of tellurium subhalides had shown these compounds to be "modified tellurium structures" (see IV), systematic investigations of their electrical conductivities were undertaken [88,90,91]. Electrical contact with subhalide crystals was achieved by alloying thin gold wires into the respective samples, and a linear current-voltage relationship was observed. Electrical conductivities were determined during a linear change of temperature with time in an inert He-gas atmosphere. The results are given in Fig. 32 together with a schematic representation of the structural units of tellurium subhalides. According to these results, a close correlation between the conductivities and crystal structures was assumed as follows: α-TeI, which corresponds to nearly isolated Te_4I_4 molecules, is the subhalide with the highest

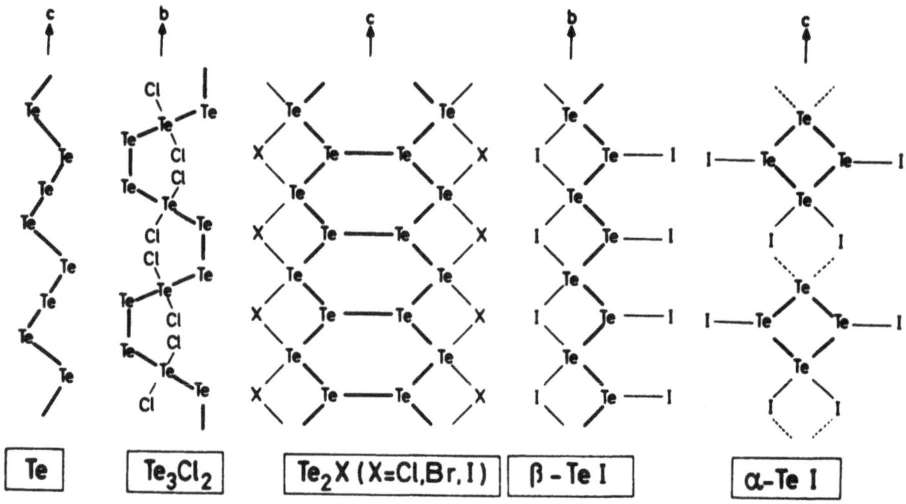

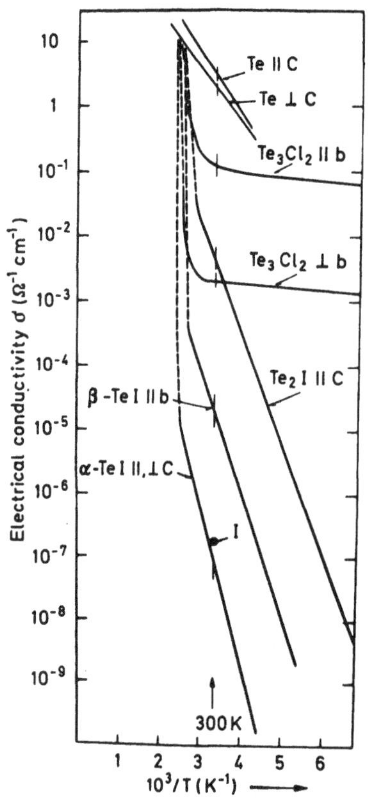

Fig. 32. Schematic representation of the structural elements of tellurium subhalides with regard to elemental Te—Te bonds above and electrical conductivity vs temperature plot of representative tellurium subhalides below [91]. The room temperature conductivity of pure iodine as well as the intrinsic conductivity lines of tellurium are indicated

resistivity. The conductivity increases in going from this nearly molecular structure to the planar zigzag chain structure of β-TeI, and again by a factor of 100 in going to the condensed chain structure of Te_2I. The highest conductivity was found for Te_3Cl_2 with a crystal structure related to elemental tellurium. The widths of the forbidden gaps of α-TeI, β-TeI and Te_2I determined by these electrical conductivity measurements were 1.0, 0.78 and 0.69 eV, respectively. As can be seen from Fig. 32, at elevated temperatures (105 °C) under the influence of current (dc), an irreversible increase of conductivity was observed. This increase was assumed to be caused by partial decomposition.

In a recent reinvestigation of the electrical conductivity [92] of the thermodynamically stable subhalides Te_3Cl_2, Te_2Br and α-TeI a gas-tight cell with an adjustable additional halogen partial pressure was used. Even at higher temperatures no decomposition of the samples was observed. Furthermore, the specific electrical conductivity of Te_3Cl_2 dropped below the respective values of α-TeI as a result of an increase of the quality of the single crystals used (see III.C).

Thus, the simple concept of a close correlation between the electrical conductivities and the subhalide structural units (see Fig. 32) was destroyed. The electrical conductivity versus temperature plots of Te_3Cl_2, Te_2Br and α-TeI, respectively, are shown in Figs. 33–35. The widths of the forbidden gaps derived from these measurements (Te_3Cl_2 1.35 eV, Te_2Br 0.6 eV, α-TeI 1.3 eV) are in good agreement with the corresponding optical investigations (Te_3Cl_3 1.3 eV [7], Te_2Br 0.6 eV [7], α-TeI 1.32 eV [50,85,86]). Significant anisotropies in the electrical conductivity detected by using different orientations of the crystals were not observed. A useful modified model related to the correlation between experimental electrical conductivities and the crystal structures is based on the concept that electronic conductivities in non-metallic solids depend mainly on the overlapping of non-bonding orbitals. Such orbitals are actually present in subhalide structures (see IV.B and V). Thus, the degree of condensation of tellurium atoms is not decisive in determining the specific electrical conductivity but it is rather the different geometries of tellurium coordination and their interatomic distances which predominate.

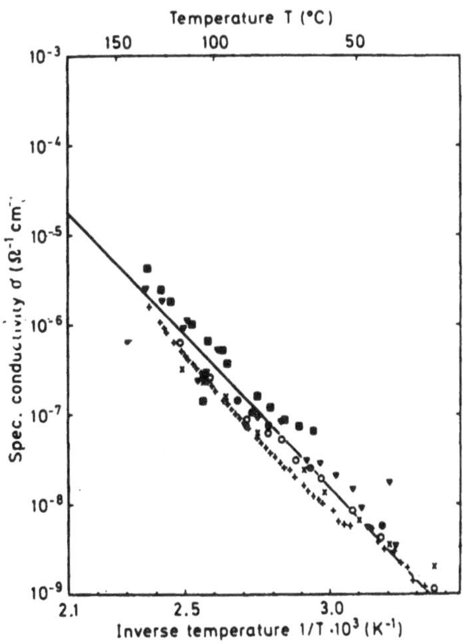

Fig. 33. Electrical conductivity vs. temperature of Te_3Cl_3; symbols refer to the corresponding number of various experiments in [92]

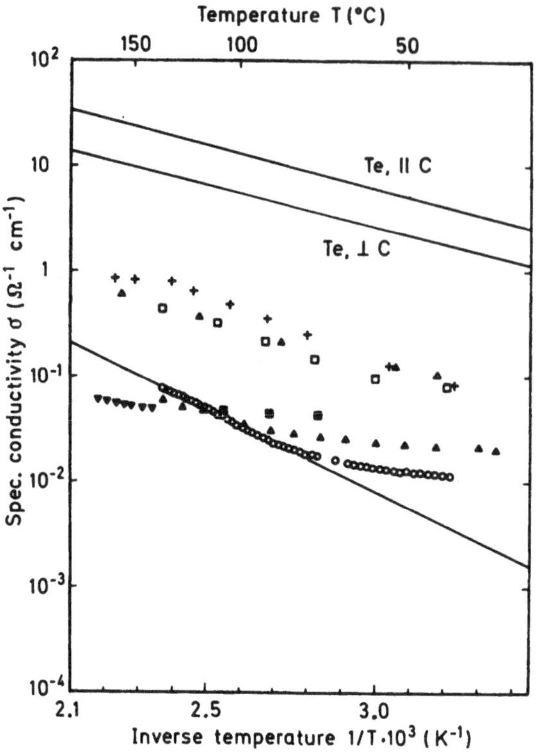

Fig. 34. Electrical conductivity vs. temperature of Te_2Br; symbols refer to the corresponding number of various experiments in [92]. The intrinsic conductivity lines of Te are indicated

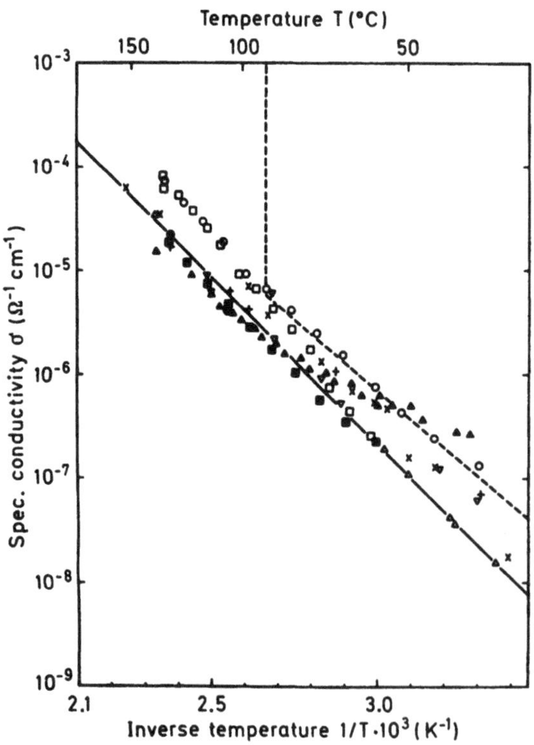

Fig. 35. Electrical conductivity vs. temperature of α-TeI; symbols refer to the corresponding number of various experiments in [92]. Dotted line: conductivity of α-TeI referring to reference [91] (see Fig. 32)

IX Solid-State Galvanic Cells

Electrical conductivity measurements using He-gas as an inert atmosphere have shown α-TeI (see VIII) to be an insulator with a room temperature conductivity comparable to that of elemental iodine. At about 375 K a steep irreversible increase in conductivity by five decades was observed. This behaviour was caused by partial decomposition of the material and accompanied by the formation of tellurium layers on its surface. These facts indicated a high diffusion rate of halogen, and solid-state galvanic cells including tellurium subhalides as solid electrolytes have been subsequently under investigation. By use of this electrochemical method data on ionic conductivity as well as thermodynamic data on the respective systems are obtained.

IX.A General Method

The first solid-state galvanic cell which was used for electromotive force (e.m.f.) measurements involving a tellurium subhalide (α-TeI)[93] is shown in Fig. 36. Cell arrangements containing various pellets of solid electrolytes in contact with Pt-electrodes at their borders were placed in an oven, the temperature of which was regulated by a programmer. The flow diagram of the complete e.m.f. measuring arrangement[93] is shown in Fig. 37. The steady-state e.m.f. of the cell is measured with a digital electrometer of extremely high input impedance ($R_i > 10^{14}\ \Omega$), the cell voltage being partially compensated by a highly constant voltage source.

In recent work[92,94] the gas-tight e.m.f. measuring cell shown in Fig. 36 was replaced by an arrangement which is presented in Fig. 38. The silver halide electrolyte is molten into a glass tube to prevent gas leakage between the silver anode and the cathode. In addition, the cathode was separated from the outer atmosphere in a gas-tight compartment to prevent evaporation of the halogen species. The free space over the cathode was made small in order to obtain steady-state conditions within reasonable time.

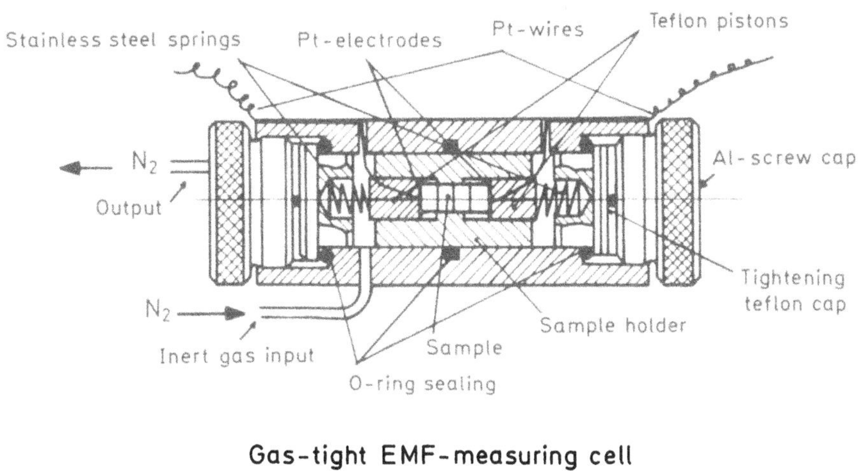

Fig. 36. Galvanic cell assembly (solid electrolytes as pellets)[93]

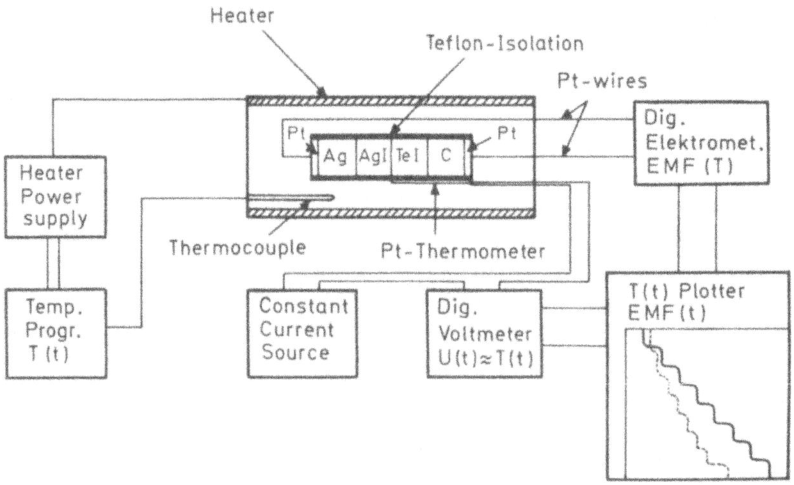

Fig. 37. Schematic representation of the complete e.m.f. measuring arrangement [93]

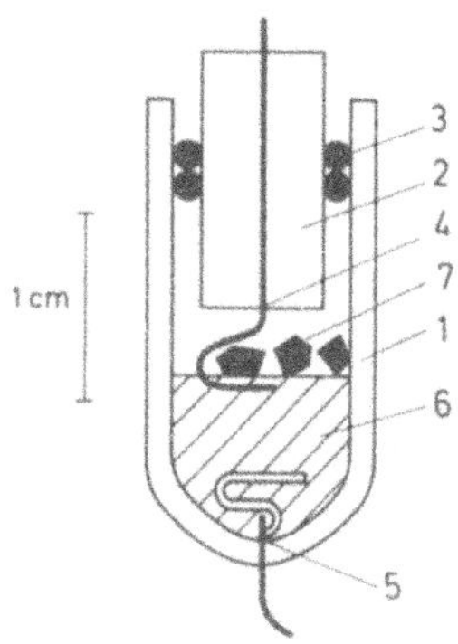

Fig. 38. Schematic representation of a modified galvanic cell assembly [92,94]: 1. outer glass tube; 2. glass piston with 0.5 mm capillary; 3. rubber rings; 4. 0.3 mm platinum wire, melted into the capillary; 5. silver wire, discharge melted to platinum wire; 6. melted-in solid electrolyte; 7. cathode mixtures

IX.B E.m.f. Measurements

The temperature dependence of the e.m.f. of cells including tellurium iodides was determined [50,93,94,95]. AgI was used as an auxiliary electrolyte; the chemical potential of α-TeI was fixed by adding Te and TeI_4, respectively (see Fig. 3).

CELL I: Pt/Ag/AgI/α-TeI, Te/C/Pt
REACTION: $Ag + \alpha\text{-}TeI \rightarrow AgI + Te$ (E_1) (11)

CELL II: Pt/Ag/AgI/α-TeI, TeI_4/C/Pt
REACTION: $3\,Ag + TeI_4 \rightarrow 3\,AgI + \alpha\text{-}TeI$ (E_2) (12)

The measurements were checked with the difference cell arrangement:

CELL III: Pt/C/α-TeI, Te/AgI/α-TeI, TeI_4/C/Pt
REACTION: $3\,Te + TeI_4 \rightarrow 4\,\alpha\text{-}TeI$ (E_3) (13)

Experimental results are shown in Fig. 39 together with the calculated curve $|E_2 - E_1|$. Crosses represent measurements with a cell arrangement without auxiliary electrolyte:

$$\text{CELL IV:} \quad \text{Pt/Ag/}\alpha\text{-TeI, Te/}\alpha\text{-TeI, TeI}_4\text{/C/Pt} \qquad (E_4) \qquad (14)$$

$$\text{REACTION:} \quad 3\,\text{Te} + \text{TeI}_4 \rightarrow 4\,\alpha\text{-TeI}$$

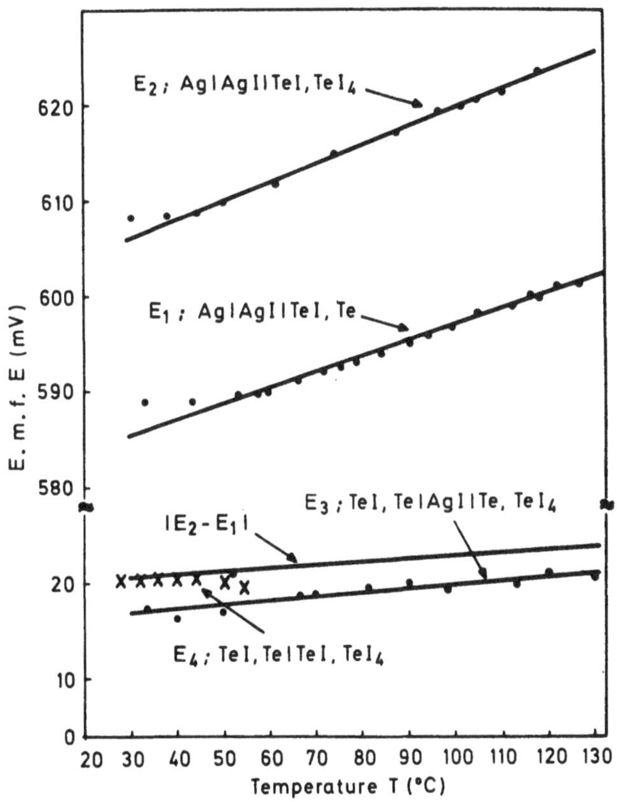

Fig. 39. Galvanic cells with α-TeI: e.m.f. vs. temperature [50)]

The good agreement with $|E_2 - E_1|$ led to the conclusion [50,93)] that α-TeI should have a remarkable partial iodine ionic conductivity. The relatively high iodine pressures of the electrode [α-TeI, TeI_4] restricted the measurements to the low temperature region.

In a recent reinvestigation with α-TeI as a solid electrolyte [92)] the theoretical e.m.f. of the difference cell E_3 was found to be shifted to higher values, which implies a decrease in the partial iodine ionic conductivity from the ionic transport number 0.8–0.9 [93)] to 0.05–0.1.

E.m.f. data as a function of temperature T may be linearized in the given temperature range using the formula (15) [96)]:

$$E = A + B \cdot T \qquad (15)$$

The coefficients A and B resulting from experimental data using galvanic cells with the stable tellurium halides [94] are listed in Table 5. The respective cell arrangements with the corresponding total cell reactions for the tellurium-iodine system have been reported ((11), (12)); the following arrangements and reactions ((16) to (19)) are appropriate to the tellurium-bromine and the tellurium-chlorine systems:

CELL V: Pt, Ag/AgBr, Te_2Br, Te, Pt
REACTION: $Te_2Br + Ag \rightarrow AgBr + 2\,Te$ (E_5) (16)

CELL VI: Pt, Ag/AgBr/C, $TeBr_4$, Te_2Br, Pt
REACTION: $2\,TeBr_4 + 7\,Ag \rightarrow 7\,AgBr + Te_2Br$ (E_6) (17)

CELL VII: Pt, Ag/AgI/AgCl, Te_3Cl_2, Te, Pt
REACTION: $Te_3Cl_2 + 2\,Ag \rightarrow 2\,AgCl + 3\,Te$ (E_7) (18)

CELL VIII: Pt, Ag/AgI/AgCl, $TeCl_4$, Te_3Cl_2, Pt
REACTION: $3\,TeCl_4 + 10\,Ag \rightarrow 10\,AgCl + Te_3Cl_2$ (E_8) (19)

Table 5. E.m.f. of cell reactions E_1, E_2 and E_{5-8} in terms of A and B of Eq. (15). Electrochemical valency, (n), temperature range and number of points used for the linear fit are indicated.

Cathode		n	A (mV)	$10^4 \times B$ (V K^{-1})	Temp.range (°C)	No. of points
E_1	α-TeI, Te	1	498.6 ± 0.6	2.11 ± 0.1	85–150	25
E_2	TeI_4, α-TeI	3	525.0 ± 0.4	1.82 ± 0.1	40–150	28
E_5	Te_2Br, Te	1	528.0 ± 0.6	1.66 ± 0.2	135–170	7
E_6	$TeBr_4$, Te_2Br	7	538.0 ± 0.7	1.24 ± 0.2	120–170	5
E_7	Te_3Cl_2, Te	2	518.0 ± 0.45	1.57 ± 0.2	60–150	6
E_8	$TeCl_4$, Te_3Cl_2	10	497.0 ± 0.1	1.6 ± 0.05	110–150	7

IX.C Thermodynamic Data

The thermodynamic data ΔG_f^0, ΔH_f^0 and S^0 of tellurium halides were calculated from the e.m.f.'s of the galvanic cells and their temperature dependences [92,94] (see IX.B) by taking into account the data for the other compounds which were involved in the total cell reactions [97]. Literature data of the standard enthalpies of formation and standard entropies of silver [98], silver iodide [99], silver chloride [98], silver bromide [100], and tellurium [99] have been used in the calculations.

Thermodynamic data on the cell reactions E_1, E_2, E_{5-8} and the data derived for the formation reactions of the tellurium halides are listed in Tables 6 and 7, respectively. A comparison of the thermodynamic data in Table 7 with the corresponding values obtained by methods other than the e.m.f. show that the determination of the e.m.f.'s is well suited to the calculation of thermodynamic functions.

The significance of the thermodynamic data for the structure and bonding conditions are discussed in the following. The temperature dependence of the experimental molar heat capacity (C_p) of α-TeI and Te_2Br as well as the low-temperature behaviour of $C_p(T)$ in terms of a C_p/T versus T^2 plot in the temperature range from

Table 6. Thermodynamic data for cell reactions E_1, E_2 and E_{5-8}.

Cathode		ΔG_r (298.15 K) (kJ mol^{-1})	ΔS_r (298.15 K) (J mol^{-1} K^{-1})	ΔH_r (298.15 K) (kJ mol^{-1})
E_1	α-TeI, Te	− 54.16 ± 0.06	20.3 ± 1	− 48.1 ± 0.4
E_2	ReI_4, α-TeI	−167.67 ± 0.12	52.7 ± 2.5	−152.0 ± 1.0
E_5	Te_2Br, Te	− 55.72 ± 0.06	16.0 ± 2.0	− 51.0 ± 0.6
E_6	$TeBr_4$, Te_2Br	−388.5 ± 0.5	83.75 ± 13.5	−363.5 ± 4.5
E_7	Te_3Cl_2, Te	−109.0 ± 0.1	30.3 ± 1.9	−100.0 ± 0.65
E_8	$TeCl_4$, Te_3Cl_2	−508.5 ± 0.1	154.85 ± 3.5	−462.3 ± 1.1

Table 7. Standard heats of formation and standard entropies of tellurium halides. Literature data obtained with other methods than e.m.f. measurements are also cited.

Compound	H_f^0 (298.15 K) (kJ mol^{-1})	S^0 (298.15 K) (J mol^{-1} K^{-1})	Method	Ref.
α-TeI	− 7.91 ± 0.4	101.6 ± 1	e.m.f.	94)
	− 7.95	118.6	vapour press.	103)
		108.7 ± 1	spec. heat	94)
TeI_4	− 42.16 ± 1	264.0 ± 2.5	e.m.f.	94)
	− 45.18	272.4	vapour press.	103)
	− 69.0 ± 12.5	226 ± 16.7	vapour press.	37)
	− 35.9 ± 8.4		heat of solut.	101)
Te_2Br	− 48.22 ± 0.6	147.53 ± 2	e.m.f.	94)
		147.8 ± 1.5	spec. heat	94)
$TeBr_4$	−189.4 ± 4.5	257.85 ± 13.5	e.m.f.	94)
	−184.48 ± 8.4		heat of solut.	101)
	−177.8 ± 1.7		heat of react.	102)
	−208.4		—	100)
Te_3Cl_2	−154.14 ± 0.65	225.55 ± 1.9	e.m.f.	94)
$TeCl_4$	−320.8 ± 1.1	202.5 ± 1.9	e.m.f.	94)
	−312.84 ± 8.4		heat of solut.	101)
	−326.35		—	99)
	−323.0		—	100)

1.5 to 5 K [94] are shown in Fig. 40. The comparatively low Debye-temperatures θ_0 for α-TeI and Te_2Br (111 and 135 K, respectively, as deduced from the insert in Fig. 40) demonstrate the rather small binding forces in these covalent compounds composed of heavy elements. The bonds are stronger in Te_2Br with a double-chain structure as compared to the molecular structure of α-TeI with the Te_4I_4-rings. A detailed inspection of the θ(T)-curve [94] of both compounds in the temperature range between 10 and 50 K did not reveal a deviation from the normal θ(T) behaviour. This is in contrast to the findings in elemental tellurium (θ_0 = 141 K), where C_p data show a constant molar heat capacity slope in the range 15 to 30 K, reflecting the typical one-dimensional character of the Te-chain [104,105]. The measurements of the subhalides suggest a less anisotropic behaviour.

Another interesting result [94] with regard to the thermodynamic properties of tellurium halides is indicated in Table 8. The standard entropies expressed per atom

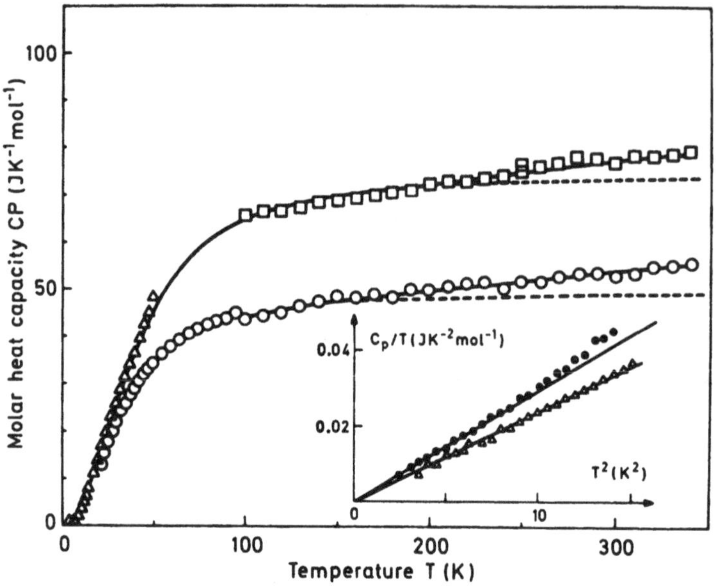

Fig. 40. Molar heat capacities of α-TeI (●, ○) and Te_2Br (△, □), measured by adiabatic calorimetry (●, △) and by differential scanning calorimetry (○, □) [94]

Table 8. Standard entropies S^0 (298.15) and standard heats of formation ΔH_f^0 (298.15) of tellurium halides per gram atom (**a**) and per halogen atom (**b**), respectively.

a	S^0 (298.15) in (J g atom^{-1} K^{-1})			
	Subhalides		Tetrahalides	
Chlorides	45.11	(Te_3Cl_2)	40.5	($TeCl_4$)
Bromides	49.2	(Te_2Br)	51.6	($TeBr_4$)
Iodides	50.6	(α-TeI)	52.8	(TeI_4)

b	ΔH_f^0 (298.15) in (kJ mol^{-1}) per halogen atom			
	Subhalides		Tetrahalides	
Chlorides	77.0	(Te_3Cl_2)	80.2	($TeCl_4$)
Bromides	48.2	(Te_2Br)	47.4	($TeBr_4$)
Iodides	7.9	(α-TeI)	10.5	(TeI_4)

are almost the same, not only within the group of subhalides and tetrahalides, but for all of the compounds, and these entropies per atom correspond to that of elemental tellurium (49.5 J atom^{-1} K^{-1} [99]). This reflects the covalent character which is common to these groups, but does not show any effects of the details of coordination, bond strengths and bond lengths. Furthermore, the standard heat of formation expressed in KJ per halogen (that is per halogen-tellurium bond) corresponds to the heat of formation of the respective halogen whether one looks at subhalides or tetrahalides.

Therefore, thermodynamics, as a macroscopic and statistic entity, seems to be less suitable to reflect details of structure and bonding as compared to spectroscopic and X-ray methods.

IX.D Vapour Pressure Determination

It was shown in [95] that e.m.f. measurements can be used to obtain accurate vapour pressure data for iodine over elemental iodine, α-TeI and TeI_4 in the range from 10^4 to 10^0 Pa and from room temperature to the transition temperature of α-AgI at 420 K. The following cell arrangements were used:

CELL IX: Ag/AgI/C or Pt, I_2 (s or l), Pt
REACTION: $Ag(s) + {}^1/_2\, I_2$ (s or l) → AgI(s) (20)

CELL X: Ag/AgI/C or Pt, α-TeI — Te, Pt
REACTION: Ag(s) + α-TeI(s) → AgI(s) + Te(s) (21)

CELL XI: Ag/AgI/C or Pt, α-TeI — TeI_4, Pt
REACTION: 3 Ag(s) + TeI_4(s) → 3 AgI(s) + α-TeI(s) (22)

The e.m.f. values as functions of the temperature of cell IX are plotted in Fig. 41 with selected literature data for comparison. The link in the e.m.f. curve clearly reflects

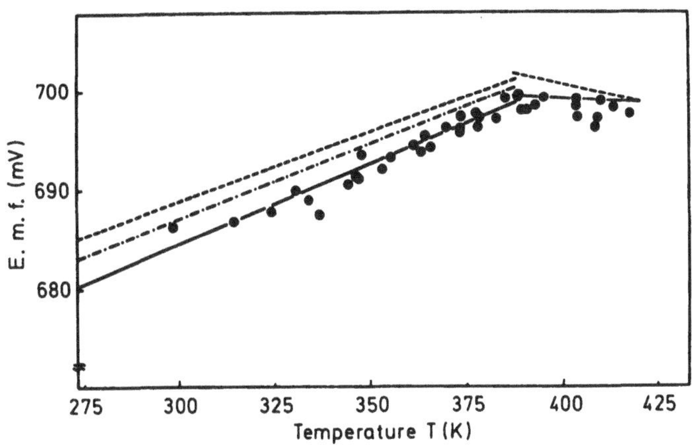

Fig. 41. e.m.f. for the AgI formation cell as a function of temperature (cell IX) [95]: ●, experimental points; ————, fitted curve to experimental points; — — — —, calculated using tabulated data from reference [99]; —·—·—·, analytical expression of the data from reference [96]

the change in Gibbs free energy for the cell reaction at the melting point of iodine. The corresponding results for cell arrangements X and XI are comparable with those given in Fig. 39. The performances of AgI formation cells with different "iodine cathodes" are compiled in Table 9.

Table 9. Performances of AgI formation cells with different "iodine cathodes".

Cathode	Temperature range (°C)	E.m.f. at 0 °C (V)	Absolute error (mV)	E/T (mV K^{-1})	Absolute error (mV K^{-1})	Ref.
$I_2(s)$	20–100	0.683	±0.9	0.15	±0.05	[96]
$I_2(s)$	25–114	0.6823	±0.4	0.146	±0.025	[95]
$I_2(l)$	114–150	0.7027	±0.4	−0.027	±0.02	[95]
CsI_3, CsI_4	40–120	0.6551	±1.1	0.175	±0.06	[96]
NH_4I, NH_4I_3	45–125	0.6442	±1.3	0.174	±0.07	[96]
RbI, RbI_3	55–130	0.629	±2.0	0.2	±0.16	[96]
CsI, CsI_3	60–150	0.6074	±2.5	0.17	±0.16	[96]
α-TeI, TeI_4	40–160	0.5747	±0.5	0.18	±0.04	[95]
Te, α-TeI	50–160	0.5561	±0.65	0.2	±0.1	[95]

The iodine partial pressures in the cathode compartments of cells X and XI were calculated using the Nernst equation for which the vapour pressure over solid iodine and its temperature dependence are given by [106]:

$$\log P_{I_2,(s)}\,(\mathrm{Pa}) = -\frac{3250}{T} + 12.475 \qquad (23)$$

E^0 was taken from the linear fit of the data obtained for cell IX. The vapour pressure calculations are compared with the corresponding measurements using the transportation method [9]. The agreement between the two methods of measuring the iodine pressure is fairly good (see Fig. 5), the deviations being almost within the reported limits of error for the transportation method.

It may be pointed out that the static e.m.f. method used for the determination of low vapour pressures of iodine seems to be superior to the dynamic transportation method.

X Conclusions and Prospective Considerations

Tellurium subhalides form a well-investigated group of compounds which can be expected to have a significant impact on current chemistry. This should be emphasized because the main interest of the binary chalcogenohalides in the fields of basic research and teaching is still centered on the tetrahalides, and in particular on the binary sulfohalides at present.

Although there have been a large number of solid-state physical investigations of tellurium subhalides, further research work on the physical properties of these compounds is to be expected. It seems to be of interest in this connection that a recent investigation [107] of superconducting transition temperatures of vapour-quenched 10000 pm-films of Te-I alloys has shown a maximum at about $Te_{50}I_{50}$ which is nearly as high as found for bulk-like films of alloys $Sn_{20}Sb_{80}$ (Fig. 42).

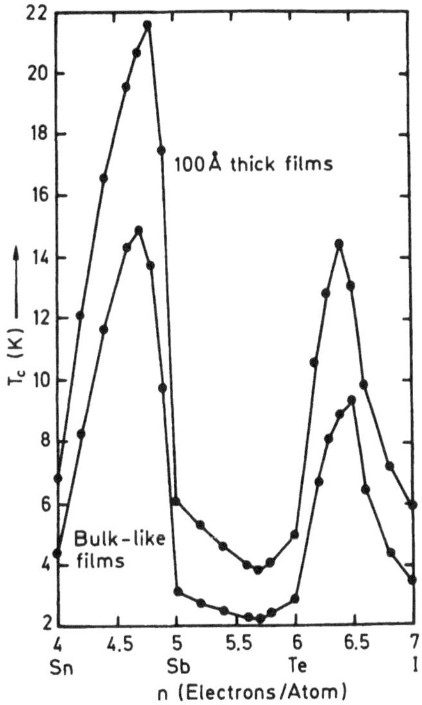

Fig. 42. Superconducting transition temperature vs. number of valence electrons per atom for 10.000 pm thick and bulk-like Al-type Sn, Sn—Sb, Sb, Sb—Te, Te, Te—I and I films [107]

Fig. 43. SEM-photograph of Te_3I crystals

From the chemical point of view it seems to be of particular interest that further (metastable) tellurium subhalides can be obtained by hydrothermal syntheses in acid solutions. In the hydrothermal work on the crystal growth of arsenic telluro-iodides [108] a pure binary tellurium subiodide with the composition Te_3I was obtained as well as a ternary phase with only small amounts of arsenic ($AsTe_9I_3$ [109]). The composition Te_3I was checked by chemical as well as energy-dispersive X-ray analysis. Crystals of the new tellurium subiodide are shown in Fig. 43; a crystal structure analysis failed because the quality of the crystals was not suitable for single crystal investigation. During annealing of the subiodide a stable phase combination α-TeI + Te was formed which indicated the metastable character of Te_3I.

XI References

1. Rose, A.: Pogg. Ann. *21*, 443 (1831)
2. Berzelius, J. J.: Ann. Chim. Phys. (1835)
3. Jaeger, F. M., Menke, J. B.: Z. Anorg. Chem. *75*, 241 (1912)
4. Damiens, A.: Bull. Soc. Chim. France *29*, 512 (1921)
5. Damiens, A.: Ann. Chim. *19*, 44 (1923)
6. Ivashin, S. A., Petrov, E. S.: Izv. Sib. Otd. Akad. Nauk SSSR, Ser. Chim. Nauk *5*, 48 (1970)
7. Rabenau, A., Rau, H.: Z. Anorg. Allg. Chem. *395*, 273 (1973)
8. Ivashin, S. A., Petrov, E. S.: Izv. Sib. Otd. Akad. Nauk SSSR, Ser. Chim. Nauk *1*, 149 (1970)
9. Kniep, R., Rabenau, A., Rau, H.: J. Less-Common Met. *35*, 325 (1974)
10. Safonov, V. V., Kuklina, N. F., Korshunov, B. G.: Russ. J. Inorg. Chem. *20*, 309 (1975)
11. Rabenau, A., Rau, H.: Proc. 6th Intern. Symp. Reactivity of Solids, Schenectady, USA, 25 (1968); (eds.) Mitchell, J. W., et al., p. 763 (1968)
12. Buss, B., Krebs, B.: Angew. Chem. *82*, 446 (1970); Angew. Chem. Int. Ed. Engl. *9*, 463 (1970)
13. Buss, B., Krebs, B.: Inorg. Chem. *10*, 2795 (1971)
14. Krebs, B., Paulat, V.: Acta Cryst. *B32*, 1470 (1976)
15. Brink Shoemaker, C., Abrahams, S. C.: Acta Cryst. *18*, 296 (1965)
16. Khodadad, P.: Ann. Chim. *3/4*, 83 (1965)
17. Blackmore, W. R., Abrahams, S. C., Kalnajs, J.: Acta Cryst. *9*, 295 (1956)
18. Kniep, R.: Thesis, Universität Braunschweig (1973)
19. Kniep, R.: in preparation
20. Prince, D. J., Corbett, J. D., Garbisch, B.: Inorg. Chem. *9*, 2731 (1970)
21. Ivashin, S. A., Petrov, E. S., Samsonova, T. I.: Izv. Sib. Otd. Akad. Nauk SSSR, Ser. Khim. Nauk *6*, 51 (1969)
22. Dharmatti, W.: Proc. Indian Acad. Sci. *12A*, 212 (1940)
23. Aynsley, E. E.: J. Chem. Soc. 3016 (1953)
24. Safonov, V. V., Konov, A. V., Korshunov, B. G.: Izv. Vysshikh. Uchebn. Zaved. Tsvetn. Met. *12*, 83 (1969)
25. Safonov, V. V., Korshunov, B. G., Pervushkin, V. V.: Izv. Vysshikh. Uchebn. Zaved. Tsvetn. Met. *13*, 97 (1970)
26. Mahmud, Y., et al.: J. Phys. Coll. *C6*, 227 (1974)
27. Wehrli, M., Gutzwiller, N.: Helv. Phys. Acta *14*, 307 (1941)
28. Voigt, A., Biltz, W.: Z. Anorg. Allg. Chem. *133*, 277 (1924)
29. Christensen, G. C., Alstad, J.: Radiochem. Radional. Lett. *13*, 227 (1973)
30. Kniep, R., Katryniok, D.: J. Chem. Soc., Dalton Trans. *1977*, 2048
31. Kniep, R.: in preparation
32. Aynsley, E. E., Watson, R. H.: J. Chem. Soc. 2603 (1955)
33. Rogers, M. T., Spurr, R. A.: J. Am. Chem. Soc. *69*, 2102 (1947)
34. Wehrli, M.: Helv. Phys. Acta *9*, 209 (1936)
35. Ivashin, S. A., Petrov, E. S.: Izv. Sib. Otd. Akad. Nauk SSSR, Ser. Khim. Nauk *2*, 28 (1971)
36. Opperman, H., Stöver, G., Wolf, E.: Z. Anorg. Allg. Chem. *410*, 179 (1974)

37. Opperman, H., Stöver, G., Wolf, E.: Z. Anorg. Allg. Chem. *419*, 200 (1976)
38. Kuliev, A. A., Kulieva, S. A., Gadzhiev, S. M.: Inorg. Mater. *8*, 717 (1972)
39. Burmeister, J.: Mater. Res. Bull. *6*, 219 (1971)
40. Grether, W.: Ann. Phys. *26*, 1 (1936)
41. Rabenau, A., Rau, H., Rosenstein, G.: Angew. Chem. *79*, 688 (1967); Angew. Chem. Int. Ed. Engl. *6*, 706 (1967)
42. Rabenau, A., Rau, H., in: Inorganic Syntheses (eds.) Wold, A., Ruff, E. K., chapter 35, New York, McGraw-Hill 1973
43. Afinogenov, Yu. P., Khaustova, Z. N.: Russ. J. Inorg. Chem. *15*, 305 (1970)
44. Kniep, R., Mootz, D., Rabenau, A.: Z. Anorg. Allg. Chem. *422*, 17 (1976)
45. Kniep, R., Mootz, D., Rabenau, A.: Angew. Chem. *85*, 504 (1973); Angew. Chem. Int. Ed. Engl. *12*, 499 (1973)
46. Rabenau, A., Rau, H., Rosenstein, G.: Angew. Chem. *82*, 811 (1970); Angew. Chem. Int. Ed. Engl. *9*, 802 (1970)
47. Kniep, R., Mootz, D., Rabenau, A.: Angew. Chem. *86*, 411 (1974); Angew. Chem. Int. Ed. Engl. *13*, 403 (1974)
48. Rabenau, A.: Hydrothermal synthesis in acid solutions, in: Crystal Growth — an Introduction (ed.) Hartmann, P., p. 198, Amsterdam, North-Holland Publ. Co. 1973
49. Rabenau, A.: High Temp.-High Press. *6*, 601 (1974)
50. Rabenau, A.: Formation and properties of some subhalides of tellurium, in: Reactivity of Solids (ed.) Wood, J., et al., p. 731, New York, Plenum Press 1977
51. Ivashin, S. A., Kanev, A. A., Petrov, E. S.: Izv. Sib. Otd. Akad. Nauk SSSR, Ser. Khim. Nauk *3*, 28 (1971)
52. Ichikawa, K., Okubo, T., Shimoji, M.: Trans. Farad. Soc. *67*, 1426 (1971)
53. Cherin, P., Unger, P.: Acta Cryst. *23*, 670 (1967)
54. Shevchik, N. J., Kniep, R.: J. Chem. Phys. *60*, 3011 (1974)
55. Ivashin, S. A., Petrov, E. S.: Izv. Sib. Otd. Akad. Nauk SSSR, Ser. Khim. Nauk *2*, 28 (1971)
56. Ivashin, S. A., Petrov, E. S.: Izv. Sib. Otd. Akad. Nauk SSSR, Ser. Khim. Nauk *5*, 34 (1970)
57. Wehrli, M.: Helv. Phys. Acta *9*, 209 (1936)
58. Oldershaw, G. A., Robinson, K.: Chem. Commun. *1970*, 540
59. Oldershaw, G. A., Robinson, K.: J. Molec. Spectrosc. *37*, 314 (1971)
60. Rieke, U.: private communication (1970)
61. Kulieva, S. A., Gadzhiev, S. M., Kuliev, A. A.: Izv. Vysshikh. Uchebn. Zaved. Chim. i Chim. Technol. *15*, 972 (1972)
62. Safonov, V. V. et al.: Russ. J. Inorg. Chem. *19*, 791 (1974)
63. Ivashin, S. A., Petrov, E. S.: Izv. Akad. Nauk SSSR, Neorg. Mater. *10*, 575 (1974)
64. Oppermann, H. et al.: Z. Anorg. Allg. Chem. *461*, 165 (1980)
65. Oldershaw, G. A., Robinson, K.: Trans. Farad. Soc. *67*, 907 (1971)
66. Bauhofer, W., Kniep, R.: Mater. Res. Bull. *8*, 989 (1973)
67. Stetter, W., Schönherr, E.: Proc. Dreiländerjahrestagung Kristallwachstum und Kristallzüchtung, p. 17, Jülich (FRG), 1975
68. Schönherr, E.: private communication 1975
69. Schönherr, E., Chen Li Chuan, Stetter, W.: private communication 1980
70. Katryniok, D., Kniep, R.: J. Therm. Anal. *16*, 291 (1979)
71. Gillespie, R. J., Nyholm, R. S.: Q. Rev. Chem. Soc. *11*, 339 (1957)
72. v. Schnering, H. G.: Angew. Chem. *93*, 44 (1981); Angew. Chem. Int. Ed. Engl. *93*, 33 (1981)
73. Schäfer, H., Eisenmann, B., Müller, W.: Angew. Chem. *85*, 742 (1973); Angew. Chem. Int. Ed. Engl. *12*, 694 (1973)
74. Pauling, L.: The Nature of the Chemical Bond, 3rd. edit., Ithaca, N.Y., Cornell University Press 1960
75. Jones, C. N. W., Mauguin, M.: J. Chem. Phys. *68*, 3067 (1978)
76. Jung, P., Triftshäuser, W.: Phys. Rev. *175*, 512 (1968)
77. Gibb, T. C. et al.: J. Chem. Soc. A. 212 (1970)
78. Takeda, M., Greenwood, N. N.: J. Chem. Soc., Dalton *1976*, 631
79. Born, P. et al.: Z. Naturforsch. *36b*, 1516 (1981)
80. Born, P., Kniep, R., Mootz, D.: Z. Anorg. Allg. Chem. *451*, 12 (1979)
81. Katryniok, D., Kniep, R.: Angew. Chem. *92*, 646 (1980); Angew. Chem. Int. Ed. Engl. *19*, 645 (1980)

82. Golubkova, B. G., Petrov, E. S.: Izv. Sib. Otd. Akad. Nauk SSSR, Ser. Khim. Nauk *2*, 114 (1975)
83. Shevchik, N. J., Cardona, M., Tejeda, J.: Phys. Rev. *B8*, 2833 (1973)
84. Kniep, R.: J. Less.-Common Met. *75*, 7 (1980)
85. v. Alpen, U.: Bull. Amer. Phys. Soc. *1975*, 495
86. v. Alpen, U., Anastasakis, E.: Verhandlg. DPG VI *10*, 375 (1975)
87. v. Alpen, U., Irslinger, C., Richter, W.: 2nd Conf. of the Condensed Matter, Div. Europ. Phys. Soc. on Dielectrics and Phonons, Budapest 1974
88. v. Alpen, U., Kniep, R.: Proc. 12th Int. Conf. on the Physics of Semiconductors, Stuttgart 1974; (ed.) Pilkuhn, M. H., p. 308, Stuttgart, Verlag Teubner 1974
89. Abrikosov, N. Kh., Zobnitsa, A. N.: V. Sb. Vopr. Met. i Fiz. Poluprov., Akad. Nauk SSSR, Moscow *1961*, 110
90. v. Alpen, U., Gmelin, E., Kniep, R.: Verhandlg. DPG VI *9*, 604 (1974)
91. v. Alpen, U., Kniep, R.: Solid-State Commun. *14*, 1033 (1974)
92. Haag, J.: Thesis, Universität Stuttgart 1978
93. v. Alpen, U., Haag, J., Rabenau, A.: Mater. Res. Bull. *11*, 793 (1976)
94. Haag, J., et al.: Z. Naturforsch. *34a*, 969 (1979)
95. Haag, J., v. Alpen, U., Rabenau, A.: J. Less-Common Met. *76*, 109 (1980)
96. Topol, L. E.: Inorg. Chem. *3*, 451 (1968)
97. Rickert, H.: Elektrochemie fester Stoffe, Berlin—Heidelberg—New York, Springer 1973
98. Codata Task Group: Recommended Key Values for Thermodynamics 1973, in: J. Chem. Thermodyn. *7*, 1 (1975)
99. Barin, B. A., Knacke, O.: Thermodynamic Properties of Inorganic Substances, Berlin—Heidelberg—New York, Springer 1973; Supplement, Berlin, Springer 1977
100. Handbook of Physics and Chemistry, Cleveland, Chemical Rubber Company 1976
101. Oppermann, H., Kunze, G., Wolf, E.: Z. Anorg. Allg. Chem. *432*, 182 (1977)
102. Tsvetkov, V. G.: Russ. J. Inorg. Chem. *23*, 1086 (1978)
103. Rau, H.: private communication 1979, as calculated from data reported in [11]
104. Walford, L. K., Carron, G. J., Schoeffel, J. A.: Mater. Res. Bull. *3*, 911 (1968)
105. de Sobo, W.: J. Chem. Phys. *21*, 764 (1953)
106. Berkenblit, M., Reismann, A.: J. Electrochem. Soc. *113*, 93 (1966)
107. Reale, C.: Phys. Lett. *74A*, 438 (1979)
108. Reski, H. D.: Thesis, Universität Düsseldorf 1981
109. Kniep, R.: unpublished